SBIR

AND THE PHASE III
CHALLENGE OF
COMMERCIALIZATION

REPORT OF A SYMPOSIUM

Committee on
Capitalizing on Science, Technology, and Innovation:
An Assessment of the Small Business Innovation Research Program

Policy and Global Affairs Division

CHARLES W. WESSNER, EDITOR

NATIONAL RESEARCH COUNCIL
OF THE NATIONAL ACADEMIES

THE NATIONAL ACADEMIES PRESS
Washington, D.C.
www.nap.edu

THE NATIONAL ACADEMIES PRESS   **500 Fifth Street, N.W.**   **Washington, DC 20001**

NOTICE: The project that is the subject of this report was approved by the Governing Board of the National Research Council, whose members are drawn from the councils of the National Academy of Sciences, the National Academy of Engineering, and the Institute of Medicine. The members of the committee responsible for the report were chosen for their special competences and with regard for appropriate balance.

This study was supported by Contract/Grant No. DASW01-02C-0039 between the National Academy of Sciences and the U.S. Department of Defense, N01-OD-4-2139 (Task Order #99) between the National Academy of Sciences and the U.S. Department of Health & Human Services, NASW-03003 between the National Academy of Sciences and the National Aeronautics and Space Administration, DE-AC02-02ER12259 between the National Academy of Sciences and the U.S. Department of Energy, and DMI-0221736 between the National Academy of Sciences and the National Science Foundation. Any opinions, findings, conclusions, or recommendations expressed in this publication are those of the author(s) and do not necessarily reflect the views of the organizations or agencies that provided support for the project.

International Standard Book Number-13:   978-0-309-10341-1
International Standard Book Number-10:   0-309-10341-X

Limited copies are available from the Policy and Global Affairs Division, National Research Council, 500 Fifth Street, N.W., Washington, DC 20001; 202-334-1529.

Additional copies of this report are available from the National Academies Press, 500 Fifth Street, N.W., Lockbox 285, Washington, DC 20055; (800) 624-6242 or (202) 334-3313 (in the Washington metropolitan area); Internet, http://www.nap.edu

Printed in the United States of America

THE NATIONAL ACADEMIES
Advisers to the Nation on Science, Engineering, and Medicine

The **National Academy of Sciences** is a private, nonprofit, self-perpetuating society of distinguished scholars engaged in scientific and engineering research, dedicated to the furtherance of science and technology and to their use for the general welfare. Upon the authority of the charter granted to it by the Congress in 1863, the Academy has a mandate that requires it to advise the federal government on scientific and technical matters. Dr. Ralph J. Cicerone is president of the National Academy of Sciences.

The **National Academy of Engineering** was established in 1964, under the charter of the National Academy of Sciences, as a parallel organization of outstanding engineers. It is autonomous in its administration and in the selection of its members, sharing with the National Academy of Sciences the responsibility for advising the federal government. The National Academy of Engineering also sponsors engineering programs aimed at meeting national needs, encourages education and research, and recognizes the superior achievements of engineers. Dr. Wm. A. Wulf is president of the National Academy of Engineering.

The **Institute of Medicine** was established in 1970 by the National Academy of Sciences to secure the services of eminent members of appropriate professions in the examination of policy matters pertaining to the health of the public. The Institute acts under the responsibility given to the National Academy of Sciences by its congressional charter to be an adviser to the federal government and, upon its own initiative, to identify issues of medical care, research, and education. Dr. Harvey V. Fineberg is president of the Institute of Medicine.

The **National Research Council** was organized by the National Academy of Sciences in 1916 to associate the broad community of science and technology with the Academy's purposes of furthering knowledge and advising the federal government. Functioning in accordance with general policies determined by the Academy, the Council has become the principal operating agency of both the National Academy of Sciences and the National Academy of Engineering in providing services to the government, the public, and the scientific and engineering communities. The Council is administered jointly by both Academies and the Institute of Medicine. Dr. Ralph J. Cicerone and Dr. Wm. A. Wulf are chair and vice chair, respectively, of the National Research Council.

www.national-academies.org

Tyrone Taylor
Director, Washington Relations
West Virginia High Technology
    Consortium Foundation

Charles Trimble
CEO (ret)
Trimble Navigation

Patrick Windham
President
Windham Consulting

PROJECT STAFF

Charles W. Wessner
Study Director

Paul Fowler
Senior Research Associate

McAlister T. Clabaugh
Program Associate

Jeffrey C. McCullough
Program Associate

David E. Dierksheide
Program Officer

Sujai J. Shivakumar
Senior Program Officer

RESEARCH TEAM

Zoltan Acs
University of Baltimore

Alan Anderson
Consultant

Philip A. Auerswald
George Mason University

Robert-Allen Baker
Vital Strategies, LLC

Robert Berger

Grant Black
University of Indiana South Bend

Peter Cahill
BRTRC, Inc.

Dirk Czarnitzki
University of Leuven

Julie Ann Elston
Oregon State University

Irwin Feller
American Association for the
    Advancement of Science

David H. Finifter
The College of William and Mary

Michael Fogarty
University of Portland

Robin Gaster
North Atlantic Research

Nicholas Karvonides

Albert N. Link
University of North Carolina

Rosalie Reugg
TIA Consulting

Donald Siegel
University of California at Riverside

Paula E. Stephan
Georgia State University

Andrew Toole
Rutgers University

Nicholas Vonortas
George Washington University

Contents

PREFACE   xiii

I. INTRODUCTION   3

II. PROCEEDINGS

Opening Remarks   33
  Charles W. Wessner, National Research Council

Introduction   37
  Jacques S. Gansler, University of Maryland

Meeting Mission Needs   44
  Charles J. Holland, Department of Defense

Panel I: The SBIR Program: Different Needs, Common Challenges   52
  Moderator: Bill Greenwalt, Senate Committee on Armed Services

  Michael Caccuitto, Department of Defense
  Michael McGrath, U.S. Navy
  Mark D. Stephen, U.S. Air Force
  John A. Parmentola, U.S. Army
  Carl G. Ray, National Aeronautics and Space Administration

Panel II: Transitioning SBIR: What Are the Issues for Prime Contractors? 75
Moderator: Max V. Kidalov, Senate Committee on Small Business and Entrepreneurship

Richard H. Hendel, Boeing Corporation
Mario Ramirez, Lockheed Martin
John P. Waszczak, Raytheon Company
Earle Rudolph, ATK

Discussant: Trevor O. Jones, BIOMEC, Inc.

Keynote Speech: Accelerating Innovation: The Luna Innovation Model 95
Kent Murphy, Luna Innovations

Panel III: Challenges of Phase III: SBIR Award Winners 102
Moderator: Kevin Wheeler, Senate Committee on Small Business and Entrepreneurship

Anthony C. Mulligan, Advanced Ceramics Research, Inc.
Nick Karangelen, Trident Systems
Thomas Crabb, Orbitec
Robert M. Pap, Accurate Automation Corporation
Mark Redding, Impact Technologies, LLC
Tom Cassin, Materials Science Corporation

Discussant: James Turner, House Committee on Science

Panel IV: Best Practice for Agency Programs: PEOs and Program Offices 124
Moderator: Peter Levine, Senate Committee on Armed Services

Richard McNamara, U.S. Navy
Stephen Lee, U.S. Army Research Office
Tracy Van Zuiden, U.S. Air Force
Peter Hughes, NASA Goddard Space Flight Center

Panel V: Lessons Learned   **140**
Moderator: Jacques S. Gansler, University of Maryland

Richard Carroll, Innovative Defense Strategies
John P. Waszczak, Raytheon Company
John Williams, U.S. Navy

Concluding Remarks   **146**
Jacques S. Gansler, University of Maryland

III. APPENDIXES

A. Biographies of Speakers   **149**

B. Participants List   **171**
   14 June 2005 Symposium

C. Bibliography   **177**

Preface

Today's knowledge economy is driven in large part by the nation's capacity to innovate. One of the defining features of the U.S. economy is a high level of entrepreneurial activity. Entrepreneurs in the United States see opportunities and are willing and able to take on risk to bring new welfare-enhancing, wealth-generating technologies to the market. Yet, while innovation in areas such as genomics, bioinformatics, and nanotechnology present new opportunities, converting these ideas into innovations for the market involves substantial challenges.[1] The American capacity for innovation can be strengthened by addressing the challenges faced by entrepreneurs. Public-private partnerships are one means to help entrepreneurs bring new ideas to market.[2]

The Small Business Innovation Research (SBIR) program is one of the largest examples of U.S. public-private partnerships. Founded in 1982, SBIR was designed to encourage small business to develop new processes and products and to provide quality research in support of the many missions of the U.S. government. By including qualified small businesses in the nation's R&D effort, SBIR grants and contracts are intended to stimulate innovative new technologies to help agencies meet the specific research and development needs of the nation in

---

[1]See Lewis M. Branscomb, Kenneth P. Morse, Michael J. Roberts, and Darin Boville, *Managing Technical Risk: Understanding Private Sector Decision Making on Early Stage Technology Based Projects,* Washington, D.C.: Department of Commerce/National Institute of Standards and Technology, 2000.

[2]For a summary analysis of best practice among U.S. public-private partnerships, see National Research Council, *Government-Industry Partnerships for the Development of New Technologies: Summary Report*, Charles W. Wessner, ed., Washington, D.C.: National Academies Press, 2002.

many areas, including health, the environment, and national defense. The SBIR program is today the largest of the government's efforts to draw on the inventiveness of small, high-technology firms, with a budget of $1.85 billion for 2005.[3]

THE NATIONAL RESEARCH COUNCIL ASSESSMENT OF SBIR

As the SBIR program approached its twentieth year of operation, the U.S. Congress asked the National Research Council (NRC) to conduct a "comprehensive study of how the SBIR program has stimulated technological innovation and used small businesses to meet federal research and development needs" and make recommendations on improvements to the program. HR 5667 directs the NRC to evaluate the quality of SBIR research and evaluate the SBIR program's value to the mission of the agencies that administer it. It calls for an assessment of the extent to which SBIR projects achieve some measure of commercialization, as well as an evaluation of the program's overall economic and non-economic benefits. It also calls for additional analysis as required to support specific recommendations on areas such as measuring outcomes to enhance agency strategy and performance, increasing Federal procurement of technologies produced by small business, and overall improvements to the SBIR program.

It is important to note that the NRC Committee assessing the SBIR program was not asked to consider if SBIR should exist or not—Congress has affirmatively decided this question on three occasions.[4] Rather, the Committee was charged with providing an empirically based assessment of the program's operations, achievements, and challenges to improve public understanding of the program and to develop recommendations to enhance the program's effectiveness.

With regard to the program's effectiveness, it became apparent in the course of the Academies' review that the Phase III element of the SBIR program would benefit from further examination. This need seemed particularly apparent for the agencies most often involved in the procurement of technologies developed using SBIR awards. Some in the SBIR community believe that this phase of the program could be improved. Some agencies seem to have adopted effective means of managing the Phase III transition. And in the course of the study, the prime contractors responsible for major systems at the Department of Defense (DoD) and the National Aeronautics and Space Administration (NASA) have shown greater interest in the SBIR program, seeing it increasingly as a wellspring of innovative technologies.

---

[3]U.S. Small Business Administration TechNet Data Base, <*http://tech-net.sba.gov/*>, Accessed on July 25, 2006.

[4]These are the 1982 Small Business Development Act and the subsequent multi-year reauthorizations of the SBIR program in 1992 and 2000.

To capture these various perspectives, the Academies convened the conference on "SBIR and the Phase III Challenge of Commercialization" on June 14, 2005. The meeting focused on the commercialization of SBIR-funded innovations at DoD and NASA, where commercialization often takes the form of agency acquisition. It was held under the leadership of Jacques Gansler, vice president for research at the University of Maryland and former Under Secretary of Defense for Acquisition, Technology and Logistics.

A unique feature of the conference is that it brought together, for the first time, the program managers, small business leaders, and prime contractor personnel involved in commercializing the results of SBIR awards through procurement at the DoD and NASA. These participants identified the challenges as well as highlighted existing and evolving best practices among successful cases in the third (or commercialization) phase of the SBIR program. This conference, summarized in this report, covered a rich variety of topics though, given the one-day timeframe of the meeting and the richness of the subject, did not (and indeed could not) cover the many possible issues associated with the program. The conference and this report do have the virtue of focusing on a key element of the SBIR program—the Phase III transition.[5]

ACKNOWLEDGMENTS

On behalf of the National Academies, we express our appreciation and recognition for the insights, experiences, and perspectives made available by the participants in the conference.

A number of individuals deserve recognition for their contributions to the preparation of the conference and this report. These include Ken Jacobson, Robin Gaster, Sujai Shivakumar, McAlister Clabaugh, and David Dierksheide. Without their collective efforts, amidst many other competing priorities, it would not have been possible to prepare this report in the required period.

NATIONAL RESEARCH COUNCIL REVIEW

This report has been reviewed in draft form by individuals chosen for their diverse perspectives and technical expertise, in accordance with procedures approved by the National Academies' Report Review Committee. The purpose of

---

[5]This conference focuses on commercialization, one of four goals of the Small Business Innovation Research (SBIR) program. Created in 1982 through the Small Business Innovation Development Act, the program is designated with four distinct purposes: "(1) to stimulate technological innovation; (2) to use small business to meet federal research and development needs; (3) to foster and encourage participation by minority and disadvantaged persons in technological innovation; and (4) to increase private sector commercialization innovations derived from Federal research and development."

this independent review is to provide candid and critical comments that will assist the institution in making its published report as sound as possible and to ensure that the report meets institutional standards for quality and objectivity. The review comments and draft manuscript remain confidential to protect the integrity of the process.

We wish to thank the following individuals for their review of this report: Robert Archibald, College of William and Mary; Robert Genco, State University of New York at Buffalo; Jere Glover, Small Business Technology Coalition; and Richard Hendel, The Boeing Company.

Although the reviewers listed above have provided many constructive comments and suggestions, they were not asked to endorse the content of the report, nor did they see the final draft before its release. The review of this report was overseen by Robert White, Carnegie Mellon University, appointed by the National Academies, he was responsible for making certain that an independent examination of this report was carried out in accordance with institutional procedures and that all review comments were carefully considered. Responsibility for the final content of this report rests entirely with the authors and the institution.

STRUCTURE

Following this preface, the report's introduction describes the challenges of early stage finance in the United States and the SBIR program as well as the particular challenges of procurement for DoD and NASA. It also summarizes the key issues from the conference. The final Proceedings section of this volume provides a detailed compilation of the presentations and discussion remarks of the various speakers at the conference.

Jacques S. Gansler   Charles W. Wessner

I

INTRODUCTION

SBIR and the Phase III
Challenge of Commercialization

Small businesses are a major driver of high-technology innovation and economic growth in the United States, generating significant employment, new markets, and high-growth industries.[1] In this era of globalization, optimizing the ability of small businesses to develop and commercialize new products is essential for U.S. competitiveness and national security. Developing better incentives to spur innovative ideas, technologies, and products—and ultimately to bring them to market—is thus a central policy challenge.

Created in 1982 through the Small Business Innovation Development Act, the Small Business Innovation Research (SBIR) program is the nation's premier innovation partnership program. SBIR offers competition-based awards to stimulate technological innovation among small private-sector businesses while providing government agencies new, cost-effective, technical and scientific so-

---

[1] A growing body of evidence, starting in the late 1970s and accelerating in the 1980s indicates that small businesses were assuming an increasingly important role in both innovation and job creation. See, for example, J.O. Flender and R.S. Morse, *The Role of New Technical Enterprise in the U.S. Economy*, Cambridge, MA: MIT Development Foundation, 1975, and David L. Birch, "Who Creates Jobs?" *The Public Interest*, 65:3-14, 1981. Evidence about the role of small businesses in the U.S. economy gained new credibility with the empirical analysis by Zoltan Acs and David Audretsch of the U.S. Small Business Innovation Data Base, which confirmed the increased importance of small firms in generating technological innovations and their growing contribution to the U.S. economy. See Zoltan Acs and David Audretsch, "Innovation in Large and Small Firms: An Empirical Analysis," *The American Economic Review*, 78(4):678-690, Sept. 1988. See also Zoltan Acs and David Audretsch, *Innovation and Small Firms*, Cambridge, MA: The MIT Press, 1990.

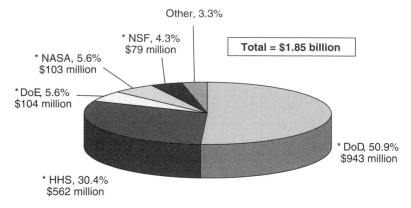

FIGURE 1 Dimensions of the SBIR program in 2005.
NOTE: These figures do not include STTR funds. * Indicates those departments and agencies reviewed by the National Research Council.
SOURCE: SBA, Retrieved July 25, 2006 from http://tech-net.sba.gov/.

lutions to meet their diverse mission needs. The program's goals are four-fold: "(1) to stimulate technological innovation; (2) to use small business to meet federal research and development needs; (3) to foster and encourage participation by minority and disadvantaged persons in technological innovation; and (4) to increase private sector commercialization derived from Federal research and development."[2]

SBIR legislation currently requires agencies with extramural R&D budgets in excess of $100 million to set aside 2.5 percent of their extramural R&D funds for SBIR. In 2005, the 11 federal agencies administering the SBIR program disbursed over $1.85 billion dollars in innovation awards. Five agencies administer over 96 percent of the program's funds. They are the Department of Defense (DoD), the Department of Health and Human Services (particularly the National Institutes of Health [NIH]), the Department of Energy (DoE), the National Aeronautics and Space Administration (NASA), and the National Science Foundation (NSF). (See Figure 1.)

As noted, a principal goal of the SBIR program is for small businesses to commercialize their innovative product or service successfully. This commer-

---

[2]The Small Business Innovation Development Act (PL 97-219). In reauthorizing the program in 1992 (PL 102-564) Congress expanded the purposes to "emphasize the program's goal of increasing private sector commercialization developed through Federal research and development and to improve the Federal government's dissemination of information concerning the small business innovation, particularly with regard to women-owned business concerns and by socially and economically disadvantaged small business concerns."

cialization can include sales to the government through public procurement as well as sales through private commercial markets. In some cases, the technology can have dual uses, with the government gaining the benefit of an innovation, which later moves into the commercial market. In other cases, the initial award successfully meets the department's goals, and no additional funds, or sales, are required.[3]

Commercializing SBIR-supported innovation is necessary if the nation is to capitalize on its SBIR investments. This transition is, however, challenging because it requires a small firm with an innovative idea to evolve quickly from a narrow focus on R&D to a much broader understanding of the complex systems and missions of federal agencies as well as the interrelated challenges of managing a larger business, developing sources of finance, and competing in the marketplace.

In cases where the federal government is the customer, small businesses must also learn to deal with a complex contracting system characterized by many arcane rules and procedures. Indeed, one major advantage of SBIR is that, to some extent, it permits small companies to sidestep some of the most impenetrable aspects of the procurement thicket that requires experienced experts in the federal acquisition regulations (FARs) to navigate and often works to the advantage of incumbents by providing the possibility of a sole source acquisition.[4] This transition to commercial or agency use is supposed to take place in the final phase (Phase III) of the SBIR program. (See Box A.)

This challenge of transition—particularly for procurement by the Department of Defense and NASA of products funded by SBIR—was the subject of an NRC conference on June 14, 2005 as one element of its congressionally requested assessment of the SBIR program (see Preface). The focus of the NRC conference was the transition of technologies from the end of SBIR Phase II into acquisition programs at the Department of Defense and NASA.

This conference report captures the informed views of conference participants but not necessarily the consensus view of the committee. This introductory chapter provides the context of the SBIR commercialization challenge and highlights some of the key points raised at the conference. The next chapter provides a detailed summary of the proceedings of the conference.

---

[3]For example, Aptima Inc. successfully completed the requirements and objectives of a Phase II contract, which developed a curriculum for teaching new Special Warfare Combatant-craft Crewmen the skills needed to maneuver boats safely in various sea conditions while maintaining speed. Aptima explicitly met the needs of the Navy on completion of Phase II, without further need of additional funds.

[4]See remarks by Kenneth Flamm in National Research Council, *SBIR: Program Diversity and Assessment Challenges,* Charles W. Wessner, ed., Washington, D.C.: The National Academies Press, 2004, pp. 10-11.

Box A SBIR—A Program in Three Phases

As conceived in the 1982 Act, the SBIR grant-making process is structured in three phases:

- **Phase I** grants essentially fund a feasibility study in which award winners undertake a limited amount of research aimed at establishing an idea's scientific and commercial promise. The 1992 legislation standardized Phase I grants at $100,000.[a]
- **Phase II** grants are larger—typically about $750,000 at DoD—and fund more extensive R&D to develop the scientific and technical merit and the feasibility of research ideas.
- **Phase III** is the period during which Phase II innovation moves from the laboratory into the marketplace. No SBIR funds support this phase. The small business must find funding in the private sector or other non-SBIR federal agency funding. To commercialize their product, small businesses are expected to garner additional funds from private investors, the capital markets, or from the agency that made the initial award. The availability of additional funds and the need to complete rigorous testing and certification requirements can pose significant challenges for new technologies and products developed under SBIR awards.

---

[a]With the agreement of the Small Business Administration, which plays an oversight role for the program, this amount can be higher in certain circumstances, e.g., drug development at NIH, and is often lower with smaller SBIR programs, e.g., the Environmental Protection Agency or the Department of Agriculture. The award levels have not been adjusted at the National Science Foundation and are therefore substantially lower today in real terms.

THE CHALLENGE OF TRANSITION AND PROCUREMENT

The challenge of technology commercialization includes the normal uncertainties of the development process common to all new technologies as well as unique institutional challenges found in federal procurement practices. Below, we first list some of the common challenges facing new firms that seek capital to develop and market their innovation. Following this, we address the specific challenges of firms that seek to commercialize their product through agency procurement.

Crossing the "Valley of Death"

Commercializing science-based innovations is inherently a high-risk endeavor.[5] One source of risk is the *lack of sufficient public information* for potential

---

[5]See, for example, Lewis M. Branscomb, Kenneth P. Morse, Michael J. Roberts, and Darin Boville, *Managing Technical Risk: Understanding Private Sector Decision Making on Early Stage*

investors about technologies developed by small firms.[6] A second related hurdle is *the leakage of new knowledge* that escapes the boundaries of firms and intellectual property protection. The creator of new knowledge can seldom fully capture the economic value of that knowledge for his or her own firm.[7]

These challenges of incomplete and leaky information pose substantial obstacles for new firms seeking capital. The difficulty of attracting investors to support an imperfectly understood, as yet-to-be-developed innovation is especially daunting. Indeed, the term, "Valley of Death," has come to describe this challenging transition when a developing technology is deemed promising, but too new to validate its commercial potential and thereby attract the capital necessary for its development.[8] Lacking the capital to develop an idea sufficiently to attract investors, many promising ideas and firms perish. (See Figure 2.[9]) Despite these challenges, many firms attempt to make their way across this Valley of Death by seeking financing from wealthy individual investors (business "angels") and, later in the development cycle, from venture capital firms.[10]

Technology Based Projects, Washington, D.C.: Department of Commerce/National Institute of Standards and Technology, 2000.

[6]Joshua Lerner, "Evaluating the Small Business Innovation Research Program: A Literature Review," in National Research Council, *The Small Business Innovation Research Program: An Assessment of the Department of Defense Fast Track Initiative,* Charles W. Wessner, ed., Washington, D.C.: National Academy Press, 2000. For a seminal analysis on information asymmetries in markets and the importance of signaling, see Michael Spence, *Market Signaling: Informational Transfer in Hiring and Related Processes,* Cambridge, MA: Harvard University Press, 1974.

[7]Edwin Mansfield, "How Fast Does New Industrial Technology Leak Out?" *Journal of Industrial Economics,* 34(2):217-224.

[8]As the September 24, 1998, Report to Congress by the House Committee on Science notes, "At the same time, the limited resources of the federal government, and thus the need for the government to focus on its irreplaceable role in funding basic research, has led to a widening gap between federally-funded basic research and industry-funded applied research and development. This gap, which has always existed but is becoming wider and deeper, has been referred to as the "Valley of Death." A number of mechanisms are needed to help to span this Valley and should be considered." See Committee on Science, *Unlocking Our Future: Toward a New National Science Policy, A Report to Congress by the House Committee on Science,* Washington, D.C.: Government Printing Office, 1998. Accessed at <*http://www.access.gpo.gov/congress/house/science/cp105-b/science105b.pdf*>.

[9]This diagram is adapted from Lewis Branscomb who in turn attributes it to a sketch made by Congressman Vernon Ehlers. See Lewis Branscomb and Philip Auswald, *Between Invention and Innovation: An Analysis of Funding for Early-Stage Technology Development,* NIST GCR 02-841, Prepared for the Economic Assessment Office, Advanced Technology Program, Gaithersburg, MD: National Institute of Standards and Technology, November 2002. For a related policy reference to the Valley of Death, see Vernon J. Ehlers, *Unlocking Our Future: Toward a New National Science Policy—A Report to Congress by the House Committee on Science,* Washington, D.C.: Government Printing Office, 1998. Accessed at <*http://www.access.gpo.gov/congress/house/science/cp105-b/science105b.pdf*>.

[10]Jeff Sohl, "The Angel Market in 2004," Center for Venture Research, University of New Hampshire. Accessed at <*http://www.unh.edu/news/docs/cvr2004.pdf*>.

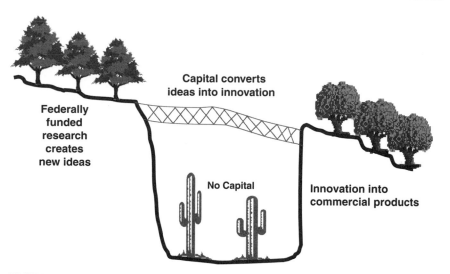

FIGURE 2 The Valley of Death

Angel investors are typically affluent individuals who provide capital for a business start-up, usually in exchange for equity. Increasingly, they organize themselves into angel networks or angel groups to share research and pool their own investment capital. The U.S. angel investment market accounted for over $22 billion in the United States in 2004.[11] It is a source of start-up capital for many new firms. Yet, the angel market is dispersed and relatively unstructured, with wide variation in investor sophistication, few industry standards and tools, and limited data on performance.[12] In addition, most angel investors are highly localized, preferring to invest in new companies that are within driving distance.[13] This geographic concentration, lack of technological focus, and the privacy concerns of many angel investors, make angel capital difficult to obtain for many high-technology start-ups, particularly those seeking to provide goods and services to the federal government.

Unlike angels, venture capitalists, typically manage the pooled money of others in a professionally-managed fund. Within the last decade, the number of venture capital firms that invest primarily in small business tripled, and their

[11]Ibid.

[12]James Geshwiler, John May, and Marianne Hudson, "State of Angel Groups," Kansas City, MO: Kauffman Foundation, April 27, 2006.

[13]See Jeffrey Sohl, John Freear, and William Wetzel, Jr., "Angles on Angels: Financing Technology-based Ventures—A Historical Perspective, *Venture Capital*, 4(4):275-287, 2002. The authors note that angel investors tend to invest close to home, "typically within a day's drive."

total investments rose eight-fold.[14] This was followed by a sharp contraction in 2000 in the venture capital market, especially for new start-ups with low valuations. A contraction in the number of initial public offerings continues to concentrate fund managers' attention on existing investments and selected "tech-bust" companies on the rebound.[15] In 2005, venture capitalists in the United States invested $21.7 billion over the course of 2,939 deals. However, 82 percent of venture capital in the United States was directed to firms in the later stages of development, with the remaining 18 percent directed to seed and early stage firms. Together, these realities of the angel and venture markets underscore the challenge faced by new firms seeking private capital to develop and market even promising innovations.

The Challenge of Federal Procurement

Commercializing SBIR-funded technologies though federal procurement is no less challenging for innovative small companies. Finding private sources of funding to further develop even successful SBIR Phase II projects—those innovations that have demonstrated technical and commercial feasibility—is often difficult because the eventual "market" for products is unlikely to be large enough to attract private venture funding. As Mark Redding of Impact Technologies noted at the conference, venture capitalists tend to avoid funding firms focused on government contracts citing higher costs, regulatory burdens, and limited markets associated with government contracting.[16]

Institutional biases in federal procurement also hinder government funding needed to transition promising SBIR technologies. Procurement rules and practices often impose high costs and administrative overheads that favor established suppliers. In addition, many acquisition officers have traditionally viewed the SBIR program as a "tax' on their R&D budgets, representing a "loss" of resources and control rather than an opportunity to develop rapid and lower cost solutions to complex procurement challenges. Even when they see the value of a technology, providing "extra" funding to exploit it in a timely manner can be a challenge that requires time, commitment, and, ultimately, the interest of those with budgetary authority for the programs or systems. Attracting such interest and support is not automatic and may often depend on personal relations and advocacy skills, not on the intrinsic quality of the SBIR project.

These acquisition hurdles and institutional bias towards SBIR remain a significant challenge for the program within DoD and NASA. Nevertheless, internal views of the SBIR program seem to be evolving in a positive fashion, although the impact of this evolution remains uneven. Some services, such as the

---

[14]Jeffrey Sohl, <*http://www.unh.edu/cvr/*>.

[15]*The Wall Street Journal*, "The Venture Capital Yard Sale," July 18, 2006, P. C1.

[16]See the presentation by Mark Redding, summarized in the Proceedings section of this volume.

Box B Technology Insertion as a Team Effort

A successful technology is the product of a complicated equation involving many different stakeholders in the acquisition community, the government S&T community and large and small businesses. "Each has a part to play, and it takes champions in each of those places to actually make it work."

Michael Caccuitto, SBIR Program Manager,
Department of Defense

Navy, have made substantial progress in changing their procurement culture so that SBIR is now seen to be a mechanism of considerable and, in some ways, unique usefulness to their acquisition officers.[17] The need to bring the relevant stakeholders together to transition a technology requires advocates and close collaboration, as aptly described in Box B.

As we see next, this change in the perception of SBIR is becoming more widespread in the DoD acquisition process.[18] In a major perceptual and political shift (that has taken place over the course of the Academies review,) prime contractors have begun to express much more interest in working with SBIR companies and are increasingly devoting management resources to capitalize on the opportunities offered by the SBIR awardees.[19] This perceptual shift is important in that it validates the program in the acquisition process while opening opportunities for firms and program managers to transition the results of successful SBIR awards into systems and products to support the DoD mission.

SBIR PHASE III: ACTIVITIES AND OPPORTUNITIES

The Department of Defense and SBIR

Reflecting this evolution in the perception of the program, Dr. Charles Holland of the Office of the Secretary of Defense (OSD) affirmed that SBIR is

---

[17]See the summaries of presentations by Deputy Assistant Secretary Michael McGrath, Navy Program Executive Officer Richard McNamara, Navy Program Manager John Williams, and Richard Carroll of Innovative Defense Strategies in the Proceedings section of this volume.

[18]See Deputy Under Secretary of Defense (Industrial Policy), "Transforming the Defense Industrial Base: A Roadmap," February 2003. This report highlights the importance of small businesses for building future war-fighting capacity. Accessed at <*http://www.acq.osd.mil/ip/docs/ transforming_the_defense_ind_base-full_report_with_appendices.pdf*>.

[19]See the presentations by Boeing, Lockheed Martin, Raytheon, and ATK in Panel II of the Proceedings section of this volume.

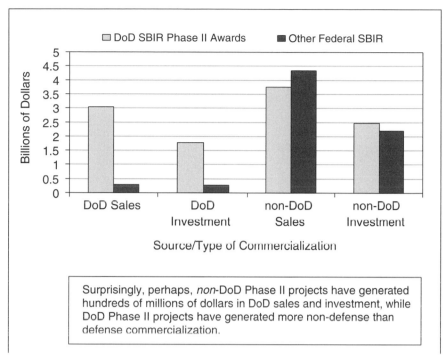

Surprisingly, perhaps, *non*-DoD Phase II projects have generated hundreds of millions of dollars in DoD sales and investment, while DoD Phase II projects have generated more non-defense than defense commercialization.

FIGURE 3 Phase III composition: Total sales and investment of DoD Phase II awards. NOTE: Sixty-three percent of projects are DoD Phase IIs, and 37 percent are other federal Phase IIs, indicating a high degree of capture of non-DoD Phase II commercialization data.
SOURCE: Department of Defense CCR, 2005.

increasingly viewed as an important mechanism for helping to expand the nation's science and technology base. He reported that while the commercialization of SBIR-developed products at DoD is split about equally between the private sector and acquisition by DoD and its prime contractors, the growth of Phase III contracts, reported through the department's Central Contractor Registration, has outpaced the growth in the SBIR budget.

The full extent of these opportunities was also explained by Dr. Michael McGrath, Deputy Assistant Secretary of the Navy, who noted in his presentation that the Navy had had substantial success in expanding Phase III funding from $50M in 2000 to $350M in 2004. (See Figure 4.)

Dr. McGrath also underscored the utility of SBIR in offering the greater flexibility and shorter time horizon needed to move technologies more quickly into acquisition:

- **Flexibility.** SBIR offers an unusual degree of execution year flexibility,

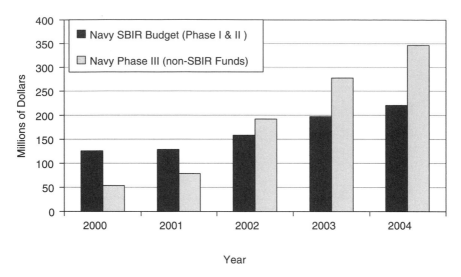

FIGURE 4 Navy Phase III success is growing.
NOTE: Phase III funding is from OSD DD 350 reports and may be under-reported. FY04 Navy Phase III ($346 million) comprises 114 separate contracts to 81 individual firms.

Box C SBIR's Value to the Department of Defense

The DoD's SBIR Program Manger, Michael Caccuitto observed that the DoD SBIR program is designed to support two key DoD objectives: creating technology dominance, and building a stronger industrial base. Drawing on this observation, Dr. McGrath affirmed that for the Navy, SBIR is an "important source of innovation across the entire RTD&E (Research, Test, Development and Evaluation) spectrum."

---

unlike most accounts for research, testing, development, and evaluation that had to be described in detail in the President's budgetary message. Dr. McGrath noted that, by contrast, the average cycle time from nominating a topic to awarding a Phase I in the Navy was 14 months.

- **Shorter Planning Horizon.** SBIR allows for a much shorter planning horizon. Most R&D programs at DoD have to be planned years ahead of the budget cycle. SBIR does not. For example, Dr. McGrath cited the case of a 2005 initiative to address the threat posed by Improvised Explosive Devices in Iraq: A quick response topic generated 38 Phase I awards within five months, 18 of which had already moved on to Phase II, with prototypes expected in the Iraq theater within six months.

Prime Contractors and SBIR

Representatives from the prime contractors indicated that SBIR Phase III collaborations are an increasing element in their corporate strategies and that these collaborations are based on growing recognition of their strategic value. These strategies include:

- **Diversifying the Supply Base.** Several prime contractors noted that they are diversifying their supplier base to avoid single points of failure. This was one factor driving their interest in the SBIR program.
- **Improving Access to Expertise.** For smaller prime contractors, SBIR also provides access to high-level expertise found within small innovative firms. As Earle Rudolph of ATK noted SBIR provides a route to new ideas and entrepreneurs able to develop them.

Richard Hendel noted that Boeing is expanding its SBIR-related activities from its Phantom Works to its Integrated Defense Systems and other large programs. Boeing has also submitted SBIR topics to agencies, some of which have been adopted. He added that DoD's Multi-mission Maritime Aircraft and Future Combat Systems programs seemed especially interested in having Boeing submit topic suggestions.

Likewise, Raytheon's John Waszczak stated that his company recognizes the value of small business collaboration. He noted that Raytheon's Integrated Defense Systems, among other corporate divisions, have been working formally with SBIR for some years. He added that half to two-thirds of a typical program for Raytheon Missile Systems have been outsourced to subcontractors, and more than half of the companies involved meet the SBA definition of small companies.

Box D Some Successful SBIR Prime Contractor-Small Business Collaborations

- Virtual Cockpit Development Program (Boeing)
- Advanced Adaptive Autopilot project, part of the Joint Direct Attack Munitions Program (JDAM) (Boeing)
- Cruise Missile Autonomous Routing System (CMARS) for the Tomahawk Mission Planning System (Boeing)
- Mark 54 Torpedo Array Nose Assembly (Raytheon)
- Exo-Atmospheric Kill Vehicle (Raytheon)

Small Businesses and SBIR

Small business speakers also provided multiple examples of successful Phase III activities that met agency needs while providing their firms with new opportunities.

In an interesting case, the head of Advanced Ceramics Research (ACR), Anthony Mulligan, described the rapid growth of his company, adding that this growth also provided much-needed employment for Native Americans at the Tohono-O'odham Reservation near Tucson, Arizona. Founded in 1989 with $1000 and the hope of a Navy Phase II award, he pointed out that his company has grown to the point where its projected revenues for 2005 were $23.5 million, of which only $4 million were from SBIR. About one-third of the company's government sales were transition dollars to get SBIR programs to commercialization. The company's Silver Fox aerial vehicle is in use in Iraq and elsewhere, providing a low-cost platform for multiple tasks.[20]

Mr. Mulligan underscored that SBIR applications at ACR are driven by the company's core strategic technology and business plan. He described his company as very disciplined in its use of the awards, noting that ACR had even turned down Phase I awards after its strategic plans had shifted away from that research topic.

The agencies, the prime contractors, and the SBIR recipient community all affirmed that with adequate Phase III funding, SBIR awards could lead to the development and delivery of important technologies that solve mission-driven problems for the agencies and the prime contractors while helping to support the growth of small high-technology businesses.

PHASE III CHALLENGES

Balancing the successes illustrated above, the conference also highlighted a wide range of concerns about Phase III from the perspectives of agency SBIR and acquisitions managers, award recipients, and prime contractors. Almost all speakers agreed that the absence of a dedicated funding source for Phase III

---

[20]A result of research and development efforts with the Navy's ONR SBIR program, the Silver Fox is a small, lightweight, inexpensive (and therefore expendable) unmanned aerial vehicle that is designed to fly autonomously for long durations at 60 knots and run on JP-5/JP-8 fuel. According to the Navy, "the Silver Fox was initially designed to spot whales in operating areas to keep them out of harm's way before conducting naval exercises. However, in 2003 the Office of Naval Research and the ACR's assembly of the Silver Fox resulted in the ability to provide operational systems to the Marine Corps who needed assistance with tactical reconnaissance missions in the Middle East. For the Marines, the Silver Fox employs high-tech "eyes" and relays information immediately to a remote laptop computer providing intelligence for advancing Marines. It has also been utilized in Operation Iraqi Freedom as an aerial chemical weapons detector." Accessed at *<http://www.dodsbir.net/SuccessStories/acr.htm>*.

Box E The Luna Innovation Model and SBIR

The Luna Innovation Model was developed by Kent Murphy who founded Luna in rural southern Virginia. The Luna model uses multiple flexible funding instruments, both public and private, including SBIR, the Advanced Technology Program (ATP), venture capital, corporate partners, and internal funding to develop and commercialize ideas that were originally generated at universities or with commercial partners.

Securing venture capital funding can be difficult even in the best of times; Luna received only two small investments during the late 1990s bubble. Also, as is noted above, venture capital firms tend to be highly specialized geographically, and Luna's southern Virginia location has minimal local venture funding.[a] The path to technical and financial success is often complex for new technologies, especially those located in more rural areas distant from high-tech clusters.

In one example, Luna Energies built its basic technology with funding from prime contractors and then used SBIR funding to develop applications for NASA and the Air Force. Eventually, it developed civilian applications for the energy industry, leading to its purchase by an energy company. According to Murphy, innovation awards from both SBIR and ATP were "critical" to Luna's success.[b]

---

[a]See Jeffrey Sohl, John Freear, and William Wetzel, Jr., "Angles on Angels: Financing Technology-based Ventures—A Historical Perspective, *Venture Capital*, 4(4):275-287, 2002.

[b]Luna Innovations is now a public company, following an initial public offering on June 9, 2006.

---

represented a key challenge. Many of them had played a part in the development of successful Phase III projects and believed that there are important lessons to be learned about how SBIR can be more successfully structured to improve technology transfer.

Agency Concerns

• **Risk Aversion by Program Officers**

Program officers and program executive officers control acquisition funds needed to move SBIR technologies eventually into weapons and other operational systems. As several conference speakers pointed out, however, Program managers and program executive officers often do not take an interest in SBIR. One reason is that while acquisition program officers are encouraged to reduce risk to the maximum extent possible, SBIR-based projects appeared to offer a number of added risks, both technical and personal, when compared to working through prime contractors.

As a result, program managers in charge of acquisitions have not traditionally seen SBIR as part of their mainstream activities. Mr. Nick Karangelen of

---

Box F DoD Risks Associated with SBIR Procurement from Small, Untested Firms

Technical Risks. This includes the possibility that the technology would not in the end prove to be sufficiently robust for use in weapons systems and space missions.

Company Risks. SBIR companies are by definition smaller and have fewer resources to draw on than prime contractors have. In addition, many SBIR companies had only a very limited track record, which limits program manager confidence that they would be able to deliver their product on time and within budget.

Funding Limitations. The $750,000 maximum for Phase II might not be enough to fund a prototype sufficiently ready for acquisition, necessitating other funds and more time.

Testing Challenges. SBIR companies are often unfamiliar with the very high level of testing and engineering specifications (mil specs) necessary to meet DoD acquisition requirements.

Scale Issues. Small companies may not have the experience and resources necessary to scale production effectively to amounts needed by DoD.

Timing Risks. DoD planning, programming, and budgets work in a two-year cycle, and it is difficult for program executive officers to determine whether a small firm will be able to create a product to meet program needs in a timely manner, even if the initial research has proven successful.

Trident Systems observed more specifically that the 100 largest contracting companies currently perform 89.9 percent of all defense R&D. Less than four percent went to small businesses. Only about 0.4 percent of all R&D generated by the government went to small technology businesses, even though one-third of all U.S. scientists and engineers were employed there.

Risk aversion is by no means peculiar to DoD. Moreover, it is entirely understandable in programs and equipment where lives could ultimately be at stake. At NASA as well, program officers usually have only one opportunity to get their projects right, given limited opportunities for in-flight adjustments. Recognizing this constraint, NASA's Carl Ray noted that many of NASA's program managers still need to be convinced that SBIR can deliver reliable technology on time and at a manageable level of risk.

• The Management Challenge: Viewing SBIR as a "tax" rather than an opportunity

As noted earlier, attitudinal issues also affect the Phase III transition, creating institutional biases in procurement that pose an extra hurdle for small firms seeking to commercialize SBIR-based products. Several speakers observed that some program managers in the acquisition offices have viewed SBIR as "more as a tax than as an opportunity" to identify and support new technologies. To them, the 2.5 percent of the agency's extramural R&D budget that is set aside for SBIR represents a loss of resources and control rather than an opportunity to develop rapid and low-cost solutions to complex procurement challenges. Illustrating this point, the Navy's McGrath noted that SBIR funds at Navy overwhelmingly came from its advanced development, testing, and evaluation functions (often referred to in the DoD idiom as 6.4-6.7 functions) but were spent on basic applied research and technology development (or 6.1-6.3 functions.) This has led to perceptions among managers involved in advanced development, testing, and evaluation that SBIR is simply a tax on their programs. This perception, in turn, can lead to limited managerial attention, less optimal mission alignment, and few resources being devoted to the program.

• **Administrative Funding Constraints**

The fact that the SBIR legislation does not permit the agencies to use SBIR funds for administration of the program is seen as another constraint. The Air Force's Major Stephen stated that four staff members at Wright-Patterson Air Force Base administer the entire Air Force SBIR program, and that while the program had experienced 70 percent growth over the previous five years, there had been no additional funding for transition assistance or program administration. In his view, the result is that the Air Force has no funds to track or document success. In turn, this made it harder to demonstrate the value of the program to acquisition program managers.

• **A Need for Gap Funding**

SBIR funding normally ends with Phase II, corresponding typically with Technology Readiness Levels (TRL) of 3-5. The SBIR program does not allow SBIR funding for further work beyond Phase II to ready the technology for use. Other acquisition funding is needed to develop the technology further. However, acquisition programs are often not prepared to fund this work given the high level of risk involved in technology development.

In Lockheed's view, the key to the TRL transition from TRL 4-5 to TRL 7-8 is the presence of available funding on hand. This reflected the comments of many speakers that smoothing the funding path across the route from TRL 4-5 to TRL 7-9 would remove a major barrier blocking improved take-up of SBIR projects into acquisition programs.

According to Mark Redding of Impact Technologies, venture capitalists were unwilling to step into the gap partly because government contracts might not be large enough to ensure the necessary level of commercial viability, and partly because the longer time horizons and significant uncertainty involved in government contacting did not fit with the relatively short time horizon and market focus of venture capital firms.

Phase III is supposed to address this gap, but many speakers noted that it had not done so very effectively. Most of the speakers discussed in some fashion the absence of dedicated funding for Phase III—in contrast to Phase II—which meant that Phase III had to be considered quite differently.

- **Cycle Time Mismatches**

In some cases, SBIR projects could be completed too soon for entry into acquisition programs that anticipate funding purchases some years out in the future. On the other hand, SBIR projects cannot be budgeted far in advance—far enough to be part of the planned acquisition program—because it was unclear whether they would be successful. These cycle time mismatches are a source of uncertainty for the program.

- **Linkages among Agencies, Prime Contractors, and Small Businesses**

Several speakers, including Kevin Wheeler of the Senate Small Business and Entrepreneurship Committee staff, noted that communication was not always good as it might be among the agencies, the prime contractors, and the small business research community. Prime contractors often have difficulty identifying the technology assets of small businesses; small businesses often had weak links to the prime contractors. The Boeing representative supported this perception, noting that the company was eager to partner with small businesses and had a significant track record with the SBIR program, but that small businesses rarely came to Boeing seeking partnerships.

- **Improving Topic Generation**

A number of speakers—including OSD's Dr. Holland and Mr. Caccuitto—said that there had been substantial improvement in the links between acquisition interests and topics, and that 60 percent or more of topics were now either sponsored or endorsed by program managers or program executive officers. In the Navy, acquisition offices supported or endorsed more than 80 percent.

However, the Air Force's Major Stephen observed that improved topic generation—i.e., development of topics more relevant to program executive officers—would also tend to reduce the timeliness of topics. Overall, the Air Force in particular believed that reducing cycle time for topic generation should be a top priority for the program.

While several speakers mentioned the need for closer links between topics and acquisition offices, John Parmentola of the Army also observed that it was necessary to balance rapid commercialization against long-term research needs. In fact, the Army needed both.

Small Businesses' Concerns

While much of the conference focused on the ways in which government agencies and prime contractors could adjust their activities to generate more effective linkages from Phase II to Phase III, several speakers observed that small businesses too could make adjustments to improve the success of Phases I, II, and III.

ACR's Anthony Mulligan said that hard work and desire were not always enough for success. He noted that there are real barriers for small businesses to overcome in linking with an acquisition program, noting that there is "no effective bridge between the acquisition community and those who are developing innovative technologies."

Many speakers supported the view that the Valley of Death between development and acquisition was a real and substantial problem for small businesses. A number of related concerns emerged:

- **Timing.** Small businesses are often disadvantaged by the very slow pace of acquisition.
- **Complexity.** The acquisition process is both complex and unique, and small firms face a steep learning curve and high overhead costs.
- **Venture Funding.** As noted, few small firms have the staff or resources to do the market analysis necessary to attract funding from venture capital firms that, in many cases, tend not to be highly motivated to invest in firms involved in government contracting.
- **Small Amount of Additional Funding.** Impact Technologies' Mark Redding noted that his company had successfully won more than 30 Phase III awards—but that these had averaged only $50,000 each.
- **Planning.** A number of agency staff noted that companies needed to be concerned with commercialization, and planning their Phase III activities, right from the start—even during Phase Zero before the first Phase I was awarded.
- **Roadmap Inclusion.** Much technical planning in acquisition is driven by roadmaps developed by program officers and prime contractors. Failure to integrate SBIR and small businesses generally into the roadmaps means that they are likely to be excluded from acquisition programs, regardless of the success of SBIR projects.
- **Contract Downsizing.** Even once a substantial Phase III has been awarded, there are no guarantees that the budget will be maintained at the contracted level. For example, Orbitec's $57 million NASA Phase III

was reduced by more than 80 percent after the first year. Such decisions can dramatically affect the transition of SBIR-funded technology.

- **Budget Squeeze.** Orbitec survived only because of successful lobbying of allies in Congress and at NASA. In general, small businesses often lack the influence to maintain budget levels when agencies change priorities—and this can be devastating for companies with few other resources to devote to a promising technology.
- **More Partnering.** A number of speakers urged small businesses to team with prime contractors rather than seek government Phase III business on their own.

Prime Contractors' Concerns

The prime contractors present at the conference also identified a number of concerns including:

- **Lack of Efficient Links to Small Firms.** Many speakers cited the example of the Navy Opportunity Forum as a means of making connections among agency program officers, SBIR program officers, prime contractors, and small businesses, as well as venture capitalists and other sources of funds.. However, the Forum was described as a unique phenomenon because other agencies do not make available the funds and management attention needed for similar activities.
- **Lack of Systemic Focus.** SBIR projects tend to focus on technical problems, not systematic needs
- **Inadequate SBIR Database for Awards and Solicitations.** Prime contractors and small firms would benefit from better capabilities for matching up prime contractor technology needs with the capacities of small firms.
- **Cultural Differences.** Differences in social and work cultures can make small businesses hesitant to work with prime contractors.
- **Lack of Evidence and Cases.** There is a need for documentation that demonstrates real positive returns on investment for the prime contractors for involving small businesses in their technology development programs.
- **Intellectual Property Concerns.** Agencies need to understand and address small business concerns about intellectual property and protect their intellectual property even under pressure to move a product forward. University partnerships are also a source of intellectual property concern for some businesses.

AGENCY INITIATIVES AND RECENT DEVELOPMENTS

In recent years, several of the DoD agencies have sought to address the issues raised above, through a variety of policy and program initiatives. Agency representatives described the initiatives undertaken in their agencies to meet these needs. These included:

- **Sponsorship of Topics.** Acquisition offices currently sponsor or endorse more than half of all DoD topics. At Navy, the acquisition-driven model of topic development had been expanded further, according to Dr. McGrath. He noted that 84 percent of Navy topics came from the acquisition community and that Program Executive Officers in the Navy's Systems Commands participated in selecting proposals and managing them through Phase I and Phase II.
- **Direct Program Executive Officer Sponsorship Pilot.** A 2005 Army pilot program to allocate 10 topics to program executive officers also had the additional effect of driving SBIR toward applied research. This constitutes a shift away from the traditional Army Research Office focus on more basic research.
- **The Navy "Primes Initiative."** Begun in 2002, the Navy Primes Initiative is an effort to connect prime contractors to the SBIR program in a more formal way. As noted, prime contractors have become increasingly interested in more access to the SBIR program.
- **Fast Track Initiative.** Started in 1995, this initiative is aimed at speeding up Phase II awards for companies that could demonstrate matching funds.[21]
- **Extra large Awards.** Larger awards (beyond $750,000) are sometimes used, partly as a way of "exciting the interest of program officers," according to Major Stephen.
- **The Transition Assistance Program (TAP).** The Navy's TAP provides mentoring and a management assistance program for supporting commercialization (i.e., transition) through the Phase III maturation process.[22]

---

[21]In an earlier assessment of the Fast Track program, the NRC noted that "the case studies, surveys, and empirical research suggest that the Fast Track initiative is meeting its goals of encouraging commercialization and attracting new firms to the program." See National Research Council, *The Small Business Innovation Research Program: An Assessment of the Department of Defense Fast Track Initiative*, Charles W. Wessner, ed., Washington, D.C.: National Academy Press, 2000.

[22]According to the Navy, the goals of the Transition Assistance Program are to facilitate DoD use of Navy-funded SBIR technology and to assist SBIR-funded firms to speed up the rate of technology transition by developing relationships with prime contractors, as well as other activities aimed at preparing the SBIR firm to deliver product. TAP is a competitive 10-month program offered exclusively to SBIR and STTR Phase II award recipients. Information accessed at <*http://www.dawnbreaker.com/navytap/*>.

- **Training and Education.** The Air Force has implemented a training and education program for prime contractors and program offices.
- **Better Tracking.** Improved outcomes tracking through the Commercialization Achievement Index (CAI) was established in 2000 to measure the commercial track record of proposing firms.
- **Outreach.** As noted, the Navy Opportunity Forum brings together SBIR firms, prime contractors, and program executive officer and program managers, offering important networking opportunities and is well received by participants.
- **New Funding Initiatives.** These include OnPoint, the Army's venture capital initiative (soon to be paralleled by NASA's Red Planet) to help technology transitions and earn a return on investment. OnPoint invests in small entrepreneurial companies including those that would otherwise not be doing business with the Army.
- **Roadmaps.** These are initiatives focused on developing joint technology maps and coordinated planning processes. They include:
  - o Navy Advanced Technology Review Board process for evaluating across programs
  - o Joint Strike Fighter Technology Advisory Board, which reviews program priorities and includes the program office, contractor team, and S&T organizations of every service partner
- **High-quality Reviewers.** Peter Hughes of NASA noted that high-quality reviews, needed to select high-quality projects, are important for the credibility of the SBIR program and for the capacity. He noted that NASA is upgrading this area of its program.

Other Recent Developments

Complementing these agency efforts, the prime contractors also noted that they are making significant efforts recently to increase their levels of involvement in SBIR. Boeing's Richard Hendel noted, for example, that his company now has a full-time SBIR liaison, up from a previous allocation of 25 percent. Some small businesses are also now more committed to working with prime contractors, with several noting the importance of the Navy's programs in this connection.

Small businesses are also adopting a wider portfolio of strategies to improve commercialization results. For example, ACR's Anthony Mulligan noted that while his company had originally sought closer connections with program officers, it was now "reaching out to the war fighter." Once the advantages of a technology or product could be demonstrated to those charged with its use, their interest could help the company to "push the middle"—the program managers and program executive officers—to move forward with Phase III financing.

PARTICIPANTS' SUGGESTIONS FOR IMPROVING PHASE III

Some speakers focused on possible changes in agency program management, including better use of incentives for managers, roadmaps, and greater matchmaking. Others focused on ways in which small businesses and the prime contractors could better align their work to improve Phase III outcomes.

Box G   A Caveat Regarding the Issues Noted Below

This report summarizes the issues raised over the course of an NRC conference on SBIR commercialization challenge. By capturing the perspectives of agency officials, prime contractors, and small business leaders, the conference has helped to inform the deliberations of the NRC committee that is reviewing the SBIR program. The views of these participants are valuable because they reflect in many cases tacit knowledge that has been gained through operational experience with the program. By recording their observations, this report captures new information concerning managerial, performance, and cultural issues and perspectives that the SBIR program must address to realize its full potential. The inclusion of these perspectives should not be seen necessarily as an endorsement of the views of participants; they do represent informed views on potential modifications and additions to the SBIR program.

Incentives for Better Program Management

Acquisition officers play a key role in moving SBIR to Phase III. They control funding allocations, making their involvement and acceptance of SBIR critical for successful technology transition. However, as several speakers noted, program executive officers and program managers often face a range of requirements including schedule and cost constraints that could be disrupted by the failure of an unproven technology. As a result, program executive officers are often understandably risk averse, wary of new unproven technology programs, including those from SBIR.

To overcome this risk aversion, appropriate incentives have to be introduced to make SBIR technologies more viable. The Navy's McGrath noted that, with appropriate incentives, program executive officers can overcome the risks that limit their use of SBIR-funded technologies. Some of the incentives described at the conference include:

- **Alignment.** Entering the SBIR company into a program with which the program executive officer was already engaged is one way to better focus SBIR projects on outcomes that directly support agency programs (and program officer) objectives. As noted by some speakers, this could allow SBIR projects to connect with Phase III activities already under way.

- **Reliability.** This involves identifying technologies that have been operationally tested and need little if any modification. This suggestion by a participant reflected widely held views that program executive officer involvement was critical in bringing SBIR technologies to the necessary readiness level.
- **Capacity.** As Dr. McGrath noted, SBIR firms need to take steps to convince program executive officers not only that the SBIR technology works, but also that the small business will be able to produce it to scale and on time.
- **Budget Integration.** Some participants noted that program executive officers needed to see that the SBIR set-aside will be used to further their own missions. This calls for building SBIR research into the work and budget of program offices. By contrast, the Air Force's program offices submit a budget based on independent cost estimates. SBIR awards are then taken as a 2.5 percent tax out of that budget.
- **Training.** Major Stephen noted that training program executive officers to help them understand how SBIR can be leveraged to realize their mission goals is necessary. However, Mr. Carroll of Innovative Defense Strategies noted that SBIR training had been part of the general program executive officer training curriculum for one year, but had since been deleted.
- **Partnering.** As described by Carl Ray, the SBIR program at NASA is forming partnerships with mission directorates aimed at enhancing "spin-in" —the take-up of SBIR technologies by NASA programs.
- **Emphasizing Opportunity.** Dr. McGrath noted that the Navy's SBIR management attempts to provide a consistent message to program executive officers and program managers—that "SBIR provides money and opportunity to fill R&D gaps in the program. Apply that money and innovation to your most urgent needs."

Roadmap Integration

The integration of subprojects, such as those funded by SBIR, into larger operational systems is a complex and long-cycle process. For this reason, some participants emphasized the need to coordinate small business activities with prime contractor project roadmaps. Lockheed's Mr. Ramirez noted that "to make successful transitions to Phase III, SBIR technologies must be integrated into an overall roadmap." Lockheed Martin uses a variety of roadmaps to that end, including both technical capability roadmaps and corporate technology roadmaps.

The Raytheon representative added that roadmaps are important because it is necessary to coordinate the technology transition process across the customer, the supply chain, and small businesses. Coordination should include advanced technology demonstrations, which could be used to integrate multiple technologies into a complex weapons system. Raytheon's Waszczak reported that his company designates a lead executive to develop a roadmap in cooperation with DoD program managers. The roadmap then allows program officers to "generate effective pull," via the leads to the prime and to smaller subcontractors.

Lockheed's Ramirez also suggested that, in the end, improvement in Phase III outcomes would require the development of a more strategic and longer-term outlook among all the participants. Better strategic vision would also allow improved alignment between programs, prime contractors, and small businesses.

Several speakers also noted that planning activities should start very early in the technology development cycle, if possible during "Phase Zero"—the stage at which topics were being developed.

Outreach and Matchmaking

Commentary at the conference also focused on the need for more events like the Navy Opportunity Forum that foster better communication as well as on the need to improve databases that share technology results across agencies. Several prime contractor representatives supported this approach. Lockheed's Ramirez, for example, noted that his company was committed to reach out more to the small business community via the Navy Opportunity Forum and other mechanisms. Raytheon's Waszczak noted that SBIR is now considered an extension of the company's R&D program, but he also noted that effective use of the program required establishing long term relationships with key small businesses, and good coordination between acquisition managers, small business, and the prime contractors. Mr. Waszczak added that while Raytheon saw Phase II as the prime contractor's key entry point into the SBIR program, the prime contractors also need to be aware of the project at the development stage.

Specific suggestions for improving these linkages included:

- **Tracking.** Mr. Karangelen said that project tracking was insufficient. Senior agency executives were required to track SBIR projects that were part of their plans and budget as technology development continued. However, except for a few officers especially in the Navy, tracking was insufficient.
- **Improved Liaison between Acquisition and SBIR Programs.** Mr. Karangelen noted that the FY1999 defense authorization act mandated designation of liaison officers by the major acquisition programs for the SBIR community, a few individuals now represented dozens of programs. He believed that a designated liaison was needed for every program.
- **More Funding for Outreach.** A number of speakers commended the Navy's Transition Assistance Program, and several suggested that funding for similar efforts be expanded.

Box H Attributes of the Navy's SUB program

The SUB Program Executive Office is widely considered to be one of the more successful Phase III programs at DoD. The program takes a number of steps to use and support SBIR as an integral part of the technology development process.

Acquisition Involvement. SBIR opportunities are advertised through a program of "active advocacy." Program managers compete to write topics to solve their problems.

Topic Vetting. Program executive officers keep track of all topics. Program managers compete in rigorous process of topic selection. SBIR contracts are considered a reward not a burden

Treating SBIR as a Program, (including follow-up and monitoring of small businesses and how to keep them alive until a customer appears). Program managers are encouraged to demonstrate commitment to a technology by paying half of the cost of a Phase II option.

Providing Acquisition Coverage, which links all SBIR awards to the agency's acquisition program.

Awarding Phase III Contracts, within the $75 M ceiling that avoids triggering complex Pentagon acquisition rules.

Brokering Connections between SBIR and the prime contractors.

Recycling unused Phase I awards, a rich source for problem solutions.

Other Possible Agency Initiatives and Strategies

- **Small Phase III Awards.** Mr. Crabb noted that these could be a key to bridging the Valley of Death. NASA for example sometimes provided a small Phase III award—perhaps enough money to fly a demonstration payload—for a technology not ready for a full Phase III.
- **Larger Phase II Awards.** Some speakers thought that larger awards would make it easier for small firms to cross the Valley of Death.
- **Unbundling Larger Phase III Awards.** One example cited at the conference was the unbundling in 1995 of a large contract to Lockheed and McDonnell Douglas for a complex life sciences module that led to Orbitec's $57 million Phase III award.
- **Redefining SBIR and Testing and Evaluation.** Some participants suggested that DoD and SBA adopt a wider view of Research, Testing, Development and Evaluation (RDT&E), so that SBIR projects could qualify for limited testing and evaluation funding. That in turn would help fund improvements in readiness level.
- **Databases.** Some speakers observed that better technology matching capabilities would be very helpful. Suggestions included development of a frequently updated technology and report database with common organizational standards across all agencies.

- **SBIR and the Critical Path.** Mr. Hughes noted that NASA was at pains to ensure that SBIR projects were not on the critical path until risk mitigation strategies were completely in place.
- **Spring-loading Phase III.** Put in place milestones that would trigger initial Phase III funding.

Aligning Small Business Strategy

A number of speakers suggested that it was critical for small businesses to get their own strategies right in approaching SBIR with a view to moving on to Phase III. While the number of Phase III awards might be small, small businesses did have options that would enhance their chances of reaching Phase III, which was, as several speakers observed, where the real pay-off for small businesses occurred.

ACR's Mulligan noted that SBIR should be tightly connected to the company's overall strategy. His company consistently rejected SBIR topic opportunities that did not meet strategic needs, and had in at least one case returned a Phase I award that no longer fit with a changed strategic plan for the company.

Speakers also noted that it was critical for small businesses to focus on gaps identified via technology roadmaps—which represented real opportunities for Phase III—and in particular on finding ways to participate in the development of roadmaps and the identification of gaps.

Impact Technologies' Redding noted that it was possible to increase the success rate of SBIR applications, which his company had done by successfully teaming with universities and prime contractors on Phase I applications (where the latter were subcontractors), and also by ensuring that customer requirements for Phase III were part of the company's strategic approach from Phase I.

Understanding customer requirements should be part of the entire project. As the Air Force's Major Van Zuiden pointed out, companies hoping to work on the Joint Strike Fighter should realize that "weight is king," and that any proposal that was heavier than an existing technology would not fly.

Prime Contractors

Speakers from the prime contractors suggested that recent increased attention to SBIR could help improve the program in several ways, including more incentives for prime contractors to work with small business. These improvements might include:

- More funding specifically for Phase III funding, although some speakers were careful to point out that this funding should not come from the existing Phase I and Phase II funding.
- Assurance that there are realistic agency plans for Phase III.

- Better understanding among the prime contractors that existing agency requirements to work with small and disadvantaged business can be met through SBIR.
- Greater appreciation of the sole-source contracting advantages that accrue to extensions of successful SBIR awards.

SUMMARY

The views of the program managers, representatives of the prime contractors, program executive officers, and small company executives captured in this conference, and summarized in the next part of this report, reveal a growing and widely based recognition that the SBIR program can play a key role in providing timely and innovative technology solutions to agency missions.

The conference served to highlight a number of common elements, some of them relatively new developments in the perception and operation of the SBIR program. For example, the meeting revealed that the leadership at the Department of Defense, prime contractors, as well as small innovative businesses see SBIR as an increasingly important tool that aligns operational incentives with broader mission goals. Senior representatives of the Department affirmed the program's role in developing innovative solutions for mission needs. Further underscoring the program's relevance, prime contractors represented at the conference stated that they have focused management attention, shifted resources, and assigned responsibilities within their own management structures to capitalize on the creativity of SBIR firms.

The meeting also highlighted how Office of Naval Research and various branches of the Navy, especially the Navy Subs Program Executive Office, have successfully leveraged the SBIR program to advance mission needs. Their experience demonstrates that senior operational support and additional funding for program management provides legitimacy and the means needed for the program to work more effectively. In addition to funding program operations, this additional support allows for the outreach and networking initiatives (such as though the Navy Forum) among other management innovations that contribute to enhanced matchmaking, commercialization, and to the higher insertion rates for the Navy SBIR program.

In short, the experience of the Navy demonstrates that the SBIR works when each of these participants recognizes program benefits and is willing to take part in facilitating the program's operations. With the right incentives and management attention, the performance and contributions of the SBIR program might be improved. What is interesting is that each of the main actors in the DoD/NASA innovation process is increasingly finding the SBIR program to be directly relevant to their interests and objectives.

To capitalize on SBIR's potential, both better information (for small companies and large prime contractors) and supportive incentives are necessary. At the

"Program managers need incentives to work with small business. Program managers in the federal acquisition community do not intentionally shun the small business community, but they have no strong incentive to embrace a new technology or process from a small business when the risk is likely to be higher."

Anthony Mulligan, Advanced Ceramics Research

NRC conference, representatives from the agencies, small businesses, as well as the prime contractors identified additional awards, closer involvement of primes in topic selection, and better follow-on Phase III funding, and integration of SBIR companies in roadmaps and other planning devices as important to successful SBIR technology transitions.

This growing recognition of the value and potential of SBIR is now changing attitudes towards the program within the acquisition community. Program managers and executives across DoD seem to be seeing the program's potential as an integral part of the development and acquisition process. SBIR is seen less as a "tax" and more as a versatile tool to rapidly transition innovative technologies that address current mission needs.

This is a welcome development since the potential of the SBIR program to support agency missions fundamentally depends on how well it is used. A major purpose of the NRC study of SBIR is to develop and share a wider understanding of this program's achievements and challenges so that its potential may be more fully realized. The suggestions made by the participants in this conference may well contribute to this objective.

Some initial progress on new policy has already occurred. As this volume goes to press, the conference it reports on has already served to bring the issue of Phase III commercialization to the attention of Congress and Executive Branch policymakers.[23]

---

[23]Following the National Academies meeting on the SBIR commercialization challenge, the Senate Committee on Small Business & Entrepreneurship proposed legislation that called for a commercialization pilot program. See Section 252 of the 2006 National Defense Authorization Act. The bill was passed in bipartisan spirit by the Senate Committee on Small Business & Entrepreneurship (SBE) under the leadership of the committee Chair, Olympia Snowe (R-ME) and Ranking Member, John Kerry (D-MA). Further reflecting the growing appreciation of the program's role and the increased focus on Phase III, Deputy Under Secretary Finley has described the SBIR program as a means of accelerating innovation and putting better equipment into the hands of the war-fighter. See Remarks by Deputy Under Secretary Finley at the Small Business Technology Coalition Conference, Washington, D.C., September 27, 2006.

II

PROCEEDINGS

Opening Remarks

Charles W. Wessner
National Research Council

Dr. Wessner welcomed participants to the National Academies and set the stage for a discussion of the Small Business Innovation Research (SBIR) program, by noting that small companies are increasingly recognized as drivers of the nation's high-technology innovation potential and economic growth. Specifically, he noted that:

- New technologies and innovations generated by small companies are a source of new markets and high-growth industries;
- Large returns to national economic capability can result from relatively small national investments in transitioning these new technologies to the market (a reality understood by economic competitors); and that
- With appropriate policy support, promising innovations can become commercial products that drive growth.

The nation, he said, faces the overarching policy challenge of developing better incentives to convert often substantial investments in research into innovative, new technologies and then to welfare-enhancing products.

Addressing the small businessmen in the audience, he emphasized that they are one of the most important sources of vibrancy and growth for the U.S. economy. In an era of globalization, he said, enhancing the ability of small business to develop and commercialize new products would be a key element of future U.S. competitiveness. One of the strengths of small firms is their flexibility—their ability to quickly shift gears if an initial idea stalls and move into something the market wants. He added that one source of strength in the U.S. economy derives from the healthy functioning of both large and small companies and the synergies they develop.

The Complex Process of Innovation

He referred to the "global myth of the linear model" that sees innovation as a one-way process flowing from basic research to applied research to development to commercialization. This model, he said, leaves out the "messy reality" of innovation and many of the surprises that lead to technological breakthroughs. In reality, innovation is a highly complex, nonlinear process that includes major overlaps between various stages of research and development, feedback loops, unexpected outcomes, and unanticipated applications.

In addition, other common myths can obscure the transition of research into socially useful products and processes. One prevalent myth assumes that markets operate under conditions of near-perfect information. In reality, market participants have less-than-perfect knowledge, especially about the ultimate value of new ideas. It is often only the entrepreneur who recognizes the potential of his or her idea.[1] Economists call this phenomenon "asymmetric information," which can cause small firms difficulty in obtaining the funding necessary to develop new approaches for the marketplace. Obtaining the funding necessary for commercial development of new technology is not inevitable because of these gaps in infrastructure and the uncertainties associated with new ideas. Markets are powerful and often efficient over time, but they need information to operate effectively. SBIR awards contribute information that helps overcome market asymmetries.

Crossing the "Valley of Death"

A related myth is that small firms with good ideas will naturally attract the venture capital they need to commercialize their products. In reality, idea-based small firms must cross an early-stage funding gap (or developmental "valley of death") to make the transition from prototype or early product to commercial success. Many people assume that because U.S. venture capital markets are broad and deep, there is no need or role for government in this process. In reality, venture capitalists themselves acknowledge that they have limited information about

---

[1]For a review of the literature on the information gaps between entrepreneurs and investors, see Joshua Lerner, "Evaluating the Small Business Innovation Research Program: A Literature Review," in National Research Council, *The Small Business Innovation Research Program: An Assessment of the Department of Defense Fast Track Initiative*, Charles W. Wessner, ed., Washington, D.C.: National Academy Press, 2000.

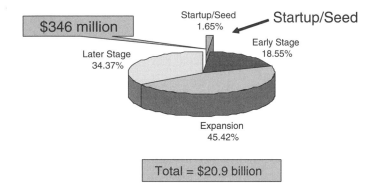

FIGURE 1 Breakdown of U.S. venture capital by stage of development, 2004.
SOURCE: PriceWaterhouseCoopers/Venture Economics/National Venture Capital Association Money Tree Survey, 2005.

new firms and tend to "follow the herd" in picking trends and products. Also, the internal dynamics of venture funds means that they focus on the latter stages of technology development. Consequently, the mainstream U.S. venture capital market tends not to focus on risky, early-stage firms, normally preferring to wait until a firm's growth prospects are well advanced before committing capital.

A breakdown of U.S. venture capital by stage of development (2004) shows that of $20.9 billion invested, only $346 million (1.65 percent) went to start-up or seed investments.[2] (See Figure 1.)

Another way of describing the funding gap encountered by small firms is to trace the path from concept to development to commercialization. At the initial single-idea level, a firm founder might begin to develop an idea for perhaps a few thousand dollars. The second step might require from $10,000 to 250,000, which is often provided by what some refer to as "friends, family, and fools." The third step in the funding ladder, requiring funds in the order of $250,000 to 500,000, may be supported by angel investors. Following this is a "missing step" where development investments in the order of $500,000 to $2 million is often required. Such investments are typically too small to interest venture capital investors but too large for most individual investors. Partial government funding of this gap in the form of competitive awards for early-stage technology development can help innovative ideas reach the market, augmenting the nation's competitiveness and security. After this stage, funding can be provided by early-stage venture capital investments ($2 to 10 million), expansion venture capital funding ($10 to 20 million), and the public equity markets.

---

[2]PriceWaterhouseCoopers/Venture Economics/National Venture Capital Association Money Tree Survey, 2005.

Examination of recent venture capital behavior reveals considerable turbulence, especially over the past decade, when total investments rose from less than $10 billion in 1995 to more than $100 billion in 2000. By 2004 investments had returned to about $20 billion. During that 2000-2004 period, while the United States experienced a five-fold decrease in available venture funding, SBIR and its sister program, the Advanced Technology Program (ATP), were in place to help address this missing step.

Focus of the Conference

The day's conference focused on Phase III SBIR activities representing product commercialization, particularly in support of the missions of the Defense Department and NASA. Specifically, the conference would:

- hear the views of small businesses, prime contractors, SBIR program officers, and government customers,
- discuss features of the program that have worked during transitions from one SBIR phase to the next, and
- identify and review features that need improvement.

He noted that the conference represented the first time the principal participants in the SBIR program—program managers, representatives of leading prime contractors, and leaders of small companies that had been successful as well as those that have faced challenges—would come together to exchange views and experiences on this topic. He expressed the hope that these views and experiences could help identify some "best practices" and/or new initiatives for the program.

The difficult challenge for the NRC committee assessing SBIR, he said, was to evaluate the program as a whole. Given SBIR's unique role, a key question in analyzing the SBIR program is, "Compared to what?" Answers to this question depend heavily on understanding the context in which SBIR operates. He concluded by emphasizing the need for policymakers to appreciate the challenges of technology transition in particular and the risks and complexity of early-stage finance in general.

Fortunately, he observed, the National Academies had assembled a distinguished committee with an exceptional chairman to take up this task. He then introduced Dr. Gansler of the University of Maryland, who is the chair of the NRC Committee evaluating SBIR.

Introduction

Jacques S. Gansler
University of Maryland

Dr. Gansler began by providing an overview of both the National Academies' SBIR evaluation study and the goals of the conference. He noted that while the Department of Defense accounted for fully half of SBIR funding, both the study and the conference were intended to examine all five major agencies with significant SBIR programs.[3]

Purpose of the SBIR Review

In summarizing the purpose of the National Academies' review, he pointed out that while the SBIR program itself was over 20 years old and had recently disbursed nearly $2 billion a year in small business funding, most reviews of the program had been largely anecdotal or internal.[4] Only recently had the first external study been done of the NIH program, which he called a "very positive step forward." He said during his time as Under Secretary of Defense, he had initiated the first comprehensive assessment of the DoD SBIR program, with the primarily purpose of evaluating the DoD Fast Track Initiative.[5]

---

[3] In 2005, DoD accounted for 50.9 percent of total SBIR dollars. Source: U.S. Small Business Administration Tech Net Data Base.

[4] In 2005, the total SBIR budget was $1,851 million. Source: U.S. Small Business Administration Tech Net Data Base.

[5] National Research Council, *The Small Business Innovation Research Program: An Assessment of the Department of Defense Fast Track Initiative*, Charles W. Wessner, ed., Washington, D.C.: National Academy Press, 2000.

That assessment recommended more external evaluations. Subsequently, Congress, in renewing SBIR in December 2000, called on the National Research Council to assess the program at the five leading SBIR agencies that together represent 96 percent of total SBIR spending.[6]

This effort did not begin immediately, partly because it required approval of all the agencies, which agreed to the study parameters only in December 2001. Funding required for work to start was received in September 2002, and the first NRC conference was held in October 2002.

Noting that he had assembled an 18-member committee to oversee the study, he added that it is an outstanding and diverse group, representing all aspects of the program, including the venture capital community, many small firms and prime contractors. This committee also oversees the work of an exceptional research team. Methodologically, the study is using a variety of approaches, including multiple surveys, interviews, and nearly 100 case studies. It would draw from a sufficiently large and representative sample of more than 4,000 firms for the Phase II survey alone.

The NRC plans to produce five kinds of reports:

- A report for Congress on program diversity and assessment challenges
- A formal methodology report required by agencies
- This report on the commercialization and Phase III challenges
- A stand-alone report on SBIR at each of the five principal agencies
- An overview of the program with findings and recommendations.

The first two reports, he said, had already been published.[7] The first report documents the wide variety of differences in the SBIR program among the agencies and even within agencies. For example, the NIH SBIR program is highly research oriented, while the DoD SBIR program is focused more on the defense mission. The chair underscored that the purpose of the current study is not to determine whether the SBIR program should continue—Congress has decided that—but to understand what could be improved, to discover what would be best practice, and to disseminate this information more widely through the agencies so as to improve the program's outcomes.

---

[6]These agencies in decreasing order of size are the Department of Defense, the National Institutes of Health, the Department of Energy, the National Aeronautics and Space Administration, and the National Science Foundation. Together they accounted in 2005 for 96.7 percent of SBIR.

[7]See National Research Council, *SBIR: Program Diversity and Assessment Challenges*, Charles W. Wessner, ed., Washington, D.C.: The National Academies Press, 2004. The Committee's methodology report is published on the web. It is available at <*http://www7.nationalacademies.org/sbir/*>.

Wide Variations Among Agencies

Because of the wide differences among agencies, individual programs used various approaches to the funding gap between Phase I and II, Phase III activities, and the length and size of awards. Because of these variations, the Committee's surveys would need to look at not just five agencies but dozens of different programs within the agencies.

He discussed the dollar values of the programs, with budget estimates for FY04 at $1.002 billion at the Department of Defense, $563 million at NIH and just over $100 million each at NASA, the Department of Energy, and the National Science Foundation (NSF). He emphasized that even though the majority of all SBIR funding was spent by just two agencies, DoD and NIH, the study would include the large number of smaller agencies that also are involved in the program, partly because those smaller agencies might be the source of ideas that would benefit other agencies. Likewise, the committee wanted to ensure that any recommendations for the large agencies would not harm the programs of the small agencies.

Dr. Gansler showed several charts illustrating the range of approaches used by different agencies, in terms of selection procedures, research topics, funding flexibility, gap funding, award cycles, and other parameters (see Figures 2, 3, and 4). It would be the NRC committee's task to identify the best practices and their most appropriate applications. A theme underlying all three figures, he said, was the program's high degree of flexibility. A challenge for the committee would be to discern how best to take advantage of that flexibility to support the mission of the SBIR program in each agency.

Program area	Range of approaches	
Selection procedures	Internal review	External review
Topics	Hard boundaries	Guidance
Funding flexibility	None	Significant
Gap funding	None	Extensive
Cycles	Annual	Multiple
Phase II plus	None	Up to $3 million
Commercialization support	None	Extensive
Phase I size	$30,000	$150,000
Phase II size	$225,000	$890,000

FIGURE 2 Significant variation among the agency SBIR programs.

	Fast Track	P1-P2 gap funding policies
NIH (FY03)	Yes	3 month extension at own risk
NSF (FY03)	No	No
DoE (FY04)	No	9 month extension P1
NASA	No	No
Navy	Yes	$30,000 and P2 selection

FIGURE 3 Variation in agency approaches to the Phase I and Phase II funding gap.
NOTE: The Navy's SBIR program is an example of the variation within the Defense Department.

NIH (FY03)	Up to $3 million	FDA-related
NSF (FY03)	$250,000	Matching funds
DoE (FY04)	No	
NASA	No	
Navy	$150,000	Agency option

FIGURE 4 Variation in agency activities beyond Phase II.
NOTE: The Navy's SBIR program is an example of the variation within the Defense Department.

He turned to the NRC assessment, which is very rigorous; subject to the National Academies' high review standards of accuracy, balance and quality; and grounded in extensive surveys and case studies. One challenge, he pointed out, is that because of the paucity of information about SBIR programs, a valid, fact-driven study—as opposed to an anecdotal or "interest-driven" study—would require new data, adding to the time required.

He reviewed in more detail the research tools that would be used in the NRC study. These include

- An extensive survey of Phase II awards for the decade 1992-2002[8]
- A Phase I award survey

---

[8]The data base stops around 2002, because considerable time is needed both to gather the needed data and to judge the impact of the program.

- A survey of program managers
- A survey of technical managers
- An extensive set of case studies.

The reason for the large anticipated size of the survey (over 100 firms will be interviewed) is that every case, in effect, is a special case. Only by gathering a large enough sample can conclusions be drawn that are more valuable than mere anecdotes. It is hoped that the case studies, together with program statistics, will lead to a deep understanding of results. He invited conference participants to submit their own suggestions for cases studies, of both successes and failures, to increase the relevant lessons learned.

The Focus on Technology Transition

He then moved to the topic of the "Phase III" challenge, saying that Congress had been urging the agencies to help small firms make the transition from the Phase II demonstration or prototype phase into an ability to commercialize or "insert" a technology into an agency acquisition program or into the public marketplace. This transition is perilous because it requires a small firm—sometimes consisting of just one or two researchers or entrepreneurs—to evolve quickly from a narrow focus on R&D to a much broader understanding of systems and missions (in the case of a federal agency) or business, finance, and competition (in the case of the public marketplace). He listed a series of measures passed by Congress that stress the desirability of Phase III activities and the need to move projects more effectively toward commercialization.[9] (See Figure 5.)

The Meaning of Commercialization

Dr. Gansler noted that an early question in the current study was whether the term "commercialization" should include sales to the government as well as sales to the private, commercial market. The answer, he said, is "yes." Many agency researchers contribute to projects for which there is little or no commercial market, such as weapons systems or space vehicles. Yet such activities should be counted as sales to a real market that fill a clear need—both important indicators of innovation and business development. One ideal outcome of the SBIR program, he said, might be a dual-use technology,[10] for the following reasons: first, the government gains the benefit of an innovation; second, if the innovation moves into the commercial market, the firm succeeds, and competition tends to drive

---

[9]The most recent of these was the 2005 National Defense Authorization Act.

[10]A dual-use technology is one that has value both for a federal agency (often defense-related) and for civilian applications.

> • **1992 SBIR Law, PL 102-564**
>   – Emphasized Commercialization (definition: "to Government or commercial markets")
> • **1999 Defense Authorization Act (Sec 818)**
>   – Emphasized push for Acquisition Offices to make Phase III awards and include SBIR in planning process
> • **1999 Senate Report 106-50 (Sec 803)**
>   – Requested that DoD develop plan to facilitate rapid transition of SBIR projects to Phase III & incorporation into DoD acquisition programs
> • **2000 SBIR Law, PL 106-554**
>   – Emphasized protection of Phase III data rights and push for more Phase III awards
> • **2005 National Defense Authorization Act (Rep.108-491)**
>   – Requires USD (AT&L) to report by March 31, 2005, information on recent Phase III awards and actions

FIGURE 5 Congress wants to increase Phase III awards and transitions into acquisition programs.

prices down and to force the technology to evolve rapidly; and third, the government as purchaser can then take advantage of the lower prices and advancing technology.

Expediting the Transition to Commercialization

He drew his own experience in DoD to recommend a three-pronged approach to assist the transition of SBIR products toward commercialization.

First, encourage program offices within the government need to play an active role. He said that in DoD, many programs took full advantage of the SBIR program by articulating their technology needs, clarifying how small businesses could help them meet these needs, and offering guidance and management assistance to small business to increase the value it delivered to the program. By contrast, he noted that many agencies in government still regarded the SBIR program as essentially a "tax" on their programs, and did not take full advantage of the SBIR program.[11] In some cases, acquisitions managers did not see the value of the program until it was explained to them.

Second, ensure that the small businesses understand and focus on technology transition and insertion. This could be accomplished through education and training of small businesses, most of which need this assistance.

---

[11]The SBIR program is a "set-aside" program funded by a 2.5 percent assessment on extramural spending for research, development, technology, and engineering.

Third, increase the involvement of prime government contractors. This could be done by creating incentives for prime contractors to "pull" SBIR technologies toward maturity, as opposed to being "pushed" ahead by a sponsoring agency. This, he said, could make a huge difference, because prime contractors (unlike most small firms) have the resources and experience to quickly bring a technology to the stage of application.

The sum of these three approaches, in his experience, could produce a significant impact on more rapid and significant commercialization and transition. A key was to develop early partnerships among the small businesses, program officers, and prime contractors in order to increase the probability of success, speed product development, reduce cost, and stimulate the defense industrial base.

He also noted additional efforts by DoD to improve SBIR outcomes—notably the Fast-Track Initiative, which provides expedited decisionmaking for SBIR awards to companies that have commitments from outside investors.[12]

An Increased Need for the SBIR Program

In summary, he said, the federal government had already had a significant impact on technology development through the SBIR program. This impact could become more important in the future as the nation's need for innovation solutions grows.

He said he had been surprised by early studies that showed how many entities were shrinking their research efforts or shifting them from basic research toward applications research and development. The federal government itself was reducing its research commitment in its 2006 budget, and other agencies were redesigning their missions to focus explicitly on development. This would have a pronounced effect on universities, which are highly dependent on the federal funding for research funding. A reduction in federal grants would increase their dependence on commercialization activities, he said, to the neglect of long-term research with high risks but high payoffs. Industry as well, he added, had reduced its research budgets in favor of an incremental and developmental focus.

These conditions, he concluded, increased the urgency to optimize the SBIR program, which is one of the few large programs available to help small, technology-based businesses survive and expand their contributions to the economy and to federal missions. The government spends about $132 billion a year on a vast range of R&D activities, only about $2 billion of which goes to the SBIR program. A careful strategy was needed if small businesses are to take full advantage of SBIR funding and to make their full contribution. The participants' help in the current evaluation, he said, would be critical in identifying those practices that best allow the private sector to capitalize on technology-based innovations.

---

[12]For a review of the Fast Track Initiative, see National Research Council, *The Small Business Innovation Research Program: An Assessment of the Department of Defense Fast Track Initiative,* Charles W. Wessner, ed., Washington, D.C.: National Academy Press, 2000.

Meeting Mission Needs

Charles J. Holland
Department of Defense

Dr. Holland said that as Deputy Under Secretary of Defense for Science and Technology, he worked for Dr. Ron Sega, the Director of Defense Research and Engineering, the chief technology officer for the Department of Defense (DoD). Dr. Holland's role was to help oversee the planning and execution of the Department's $10 to 12 billion of non-SBIR science and technology programs and to play an additional role in the SBIR program.[13]

He seconded Dr. Gansler's observation that the science and technology activities in the Department were widely distributed among the services and agencies. His job was to ensure that those activities were coordinated as parts of an integrated whole. The SBIR program was executed by every activity that had its own budget, with shared responsibility in oversight. The program also worked in partnership with the Small and Disadvantaged Business Utilization Office, managed by Frank Ramos. The DoD releases multiple solicitations, all electronically, with proposal topic generation and proposal topic review vetted by his office for technical quality and clarity.[14]

---

[13]Details of the DoD SBIR program are displayed at *<http://www.acq.osd.mil/sadbu/sbir/homepg.htm>*.

[14]By custom, government SBIR agencies use the term "topic" to refer to research topic, especially when advertising an R&D need through the SBIR program. For example, a topic listed by the DoD SBIR program as of October 2005 was "Hyperspectral/Multispectral imaging for transient events," seeking a laser-based system that could discriminate among various targets. The announcement, on the SBIR website, listed expected accomplishments for Phases I, II, and III, and also described the Private Sector Commercial Potential, which included remote sensing for various purposes and other research activities. Access at *<http://www.dodsbir.net/Topics/Default.asp>*.

Science and Technology in the DoD

He said that one reason he enjoyed working for the DoD was "that you really have a mission. People's lives are on the line, today and tomorrow, and our job is to try to have that appropriate balance of science and technology that will apply in the near term as well as the far term." He listed some "revolutionary advances" made possible by science and technology research, including stealth aircraft, adaptive optics and lasers, night vision, global positioning systems, and phased array radar. Added to this list, he said, should be the development of human-computer interfaces and much of the information technology that was critical to advanced modern defensive capabilities. He added that the development of stealth technologies that began in the 1970s were outcomes of the Cold War realization that U.S. forces could not win wars based on numbers of troops, so it would have to rely on advanced technology to maintain superiority.

He said that the defense mission would continue to grow in complexity and described three perspectives of the future. All would depend on network-centric-enabled operation, given that "we would not know where the next threat would arise." This would require networking and rapid mobility to better deal with uncertainties. Guidance would come from the White House to the Joint Chiefs of Staff.

One perspective included four kinds of operations that would be necessary for the newly complex world of the future. These include Combat Operations, which is today a strength. Two others are Stability Operations and Homeland Security, which were still challenges. Finally, Strategic Deterrence would involve communicating the message to potential enemies that the Department would be able to respond to aggression.

A second perspective of the future includes eight functional concepts required to execute operations. These included battle-space awareness, force application, command and control, focused logistics, protection, net-centric operation, joint training, and force management. His mission in science and technology was to provide knowledge and tools to support those functional concepts.

The third perspective, somewhat more complex, involves "what we've got to worry about." This list included not only about the traditional kinds of battles familiar today, but "irregular" battles that include unconventional methods adopted by non-state and state actors, terrorism, insurgency, civil war, and emerging concepts. This, he said, is the nature of every conflict today where is it not possible to draw clear lines between friend and foe.

Another challenge included "catastrophic" situations that included the use of weapons of mass destruction (WMD), a threat that is spreading throughout the world. Finally, he listed the challenge of trying to make sure that the United States is not surprised by "disruptive," previously unknown technologies or new uses of existing technologies. This, for the DoD, he said, was a challenging spectrum which demands the most effective application of science and technology,

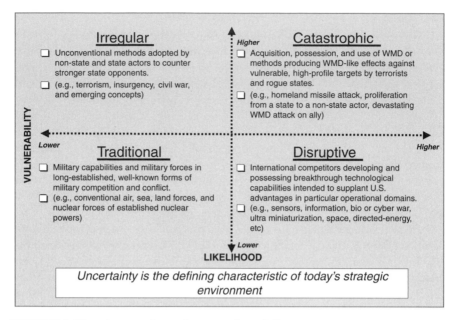

FIGURE 6 Changing security environment: four challenges.

including the development of technologies through the SBIR program. He defined the main characteristic of today's strategic environment as "uncertainty."

DoD's Strategic Plans

The DoD was approaching these challenges by coordinating a set of strategies and plans among the various military services and defense agencies. The purpose of the plans was to make sure the department, including the SBIR program, is spending its budget as effectively as it can. These strategies and plans for the Basic Research Plan (6.1) and the Defense Technology Area Plan (6.2, 6.3) are published bi-annually for Congress, DoD contractors, and others. The plans were classified, even though its individual pieces were not, because the document as a whole would constitute for an enemy a valuable insight into the DoD's intentions.

The first plan he described was the annual Joint Warfighting S&T Plan (JWSTP) to bring the power of emerging technologies to the needs of the warfighter. This plan was coordinated with the Joint Staff and the Pentagon to design science and technology programs that best support the eight functional concepts described above. Within the JWSTP are programs with milestones and metrics. He mentioned the Military Operations in Urban Terrain as an example of a defense technology program that is nearing completion. Some of the goals of the program are enhanced situational awareness, precise position location inside buildings, a combat identification system effective in buildings, improved individual mobility, and increased lethality and weapons capabilities. All of these goals required S&T components.

A second plan was the Defense Technology Area Plan (DTAP), which was "a detailed plan focusing DoD science on militarily significant technologies in specific functional areas." DTAP included near-term, applied research to enable new capabilities, such as the development of an improved jet fighter propulsion system.

A third plan is the Basic Research Plan, which supports long-term development. Each service had its own basic research program, supported by a small, service-wide program called the Defense Advanced Research Projects Agency (DARPA), originally designed to give researchers the freedom to explore potentially revolutionary ideas.[15]

DoD's S&T Strategy

Cutting across all three areas and embracing all the services and agencies, the DoD also had an S&T strategy that emphasized three areas: energy and power technologies, surveillance and knowledge systems, and the national aerospace initiative.

The first area, energy and power technologies, included many systems and ideas, in three categories. Power generation included fuel cells, fuel reforming, and novel power; energy storage developed batteries and capacitors; and power management and control developed switching and conditioning, power transmission and distribution, and thermal management. Examples of specific goals were fighter aircraft with laser weapons and electric warships with the flexibility to either use all power for steaming or to shift the electric energy for weapons use. The Army, for its future combat system, wanted vehicles using electric power, which could both propel vehicles or stop and use that energy for other purposes. He said that the DoD placed high value on these cross-cutting activities because they often led to enabling technologies and innovations. As a result, the department had doubled its investments in these projects, many of which involved small businesses, such as the development of batteries, fuel cells, and other technologies.

---

[15]Historically, DARPA has been seen as a model for how the government can foster transformative research. (National Research Council, *Assessment of Department of Defense Basic Research,* Washington, D.C.: The National Academies Press, 2005, p. 2). Yet many observers have noted that DARPA's approach has shifted away from risky research toward a focus on short-term deliverables. ("An Endless Frontier Postponed [Editorial]," *Science,* May 6, 2005; "Pentagon Redirects Its Research Dollars," *The New York Times,* April 2, 2005.

The second cross-cutting area was surveillance and knowledge systems, whose goal was to develop ideas for network-centric-enabled operations. The objectives were persistent surveillance and total knowledge of events on the battlefield, giving the ability to rapidly deploy force before the enemy could act. These systems included high-bandwidth communications, sensors and unmanned vehicles, knowledge management systems, and cyber warfare. Because these projects are common knowledge, they had to include information assurance and other protective activities.

The third cross-cutting area was the national aerospace initiative. One objective is to develop hypersonic, suborbital vehicles that can reach targets more quickly than existing craft. A second is to gain access to space from locations other than the two existing ground launching sites that might be vulnerable to attack. Techniques included two-stage-to-orbit and single-stage-to-orbit systems. Finally, the aerospace initiative is developing new space technologies, including micro-satellites and multifunction satellites.

Expanding the S&T Base

In addition to these specific programs, he described a broader effort to expand the science and technology base by using technology from other initiatives and partners. These include the Small Business Innovation Research program and its smaller cousin, the Small Business Technology Transfer (STTR) program. A third broad effort is to ensure a supply of talented people for the future through the National Defense Education Act and variants of that act. The DoD also works with both domestic and international partners to develop technology, and with other agencies, such as the National Science Foundation and the Department of Energy.

He then turned to the DoD's SBIR program, the largest source of early-stage technology financing in the United States. Of the $2 billion in total federal SBIR/STTR funding in FY05, involving 11 federal agencies, approximately half that amount will be spent by the DoD.[16] (See Figure 7.) The reason DoD has the

---

[16]The Small Business Technology Transfer (STTR) program expands public-private sector partnerships to include the joint venture opportunities for small business and the nation's premier non-profit research institutions. Federal agencies with extramural R&D budgets over $1 billion are required to administer STTR programs using an annual set-aside of 0.30 percent. Currently, five Federal agencies participate in the STTR program: DoD, DoE, DHHS (NIH), NASA, and NSF. The SBIR and STTR programs differ in two major ways. First, under the SBIR program, the Principal Investigator must have his/her primary employment with the small business concern at the time of award and for the duration of the project period. However, under the STTR program, primary employment is not stipulated. Second, the STTR program requires research partners at universities and other non-profit research institutions to have a formal collaborative relationship with the small business concern. At least 40 percent of the STTR research project is to be conducted by the small business concern, and at least 30 percent of the work is to be conducted by the single, "partnering" research institution.

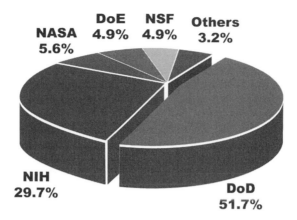

FIGURE 7 SBIR is the largest source of early-stage technology financing in the United States: SBIR/STTR agency funding.

largest SBIR program (nearly $1 billion in SBIR funding and $125 million in STTR funding) is that it taxes the entire research, development, technology and engineering (RDT&E) spectrum, not just R&D, and it includes 10 participant DoD components.[17]

The current overarching goal of the SBIR program is commercialization, and Dr. Holland reported that the DoD does "a pretty good job" in reaching this objective.

He said that 55 percent of Phase II projects were in the DoD's data base that tracks information on the commercialization success of small firms. Some 37 percent had sales and 40 percent had investment in activities that were not SBIR investments. The major sources of income were almost equal, with the private sector contributing 47 percent and the DoD or DoD prime contractors contributing 44 percent. (See Figure 8.)

As examples, he offered summaries of two of DoD's largest Phase III commercialization outcomes. The first resulted in about $763 million in sales for early work on excimer lasers, including "tools and techniques of great benefit to the semiconductor business."[18] The second one, resulting in income of several hundred million dollars, was a program that developed guidance and sensing tech-

---

[17]These are the Army, Navy, Air Force, DARPA, MDA, DTRA, SOCOM, OSD, CBD, and NGA.

[18]According to the DoD SBIR "success stories" website, "Under four DoD and DoE SBIR awards between 1989 and 1993, Science Research Laboratory, Inc. (SRL) of Somerville, Massachusetts developed a cluster of solid-state pulsed power technologies that made excimer lasers, for the first time, a commercially-viable tool for the UV lithography now used in writing current-generation integrated circuits onto computer chips." See *<http://www.acq.osd.mil/sadbu/sbir/success/index.htm>*.

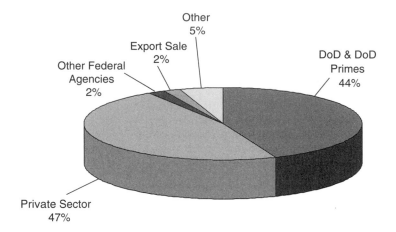

ALL DOD PHASE II PROJECTS
55% of Phase II projects in Company Commercialization Report database have
commercialization success:
37% have sales
40% have investment

FIGURE 8 DoD SBIR Phase III commercialization.
NOTE: Reported to DoD in CCR Database by firms submitting proposals to DoD in 2000-2005.

nology for anti-radiation missiles.[19] He emphasized how different were the outcomes and markets for these two SBIR projects and noted that such diversity characterized SBIR activities in general and contributed to its success.

Echoing a point made earlier by Dr. Gansler, he said that the SBIR program produced not only large systems for commercialization, but also small components and important incremental improvements. He showed a slide of such small successes produced for the Army that would be used in the Iraqi theater, including:

- Components for miniature portable power supplies developed by Mesoscopic Devices
- A Shot Pocket Charger technology developed by Space Hardware Optimization Technology
- Cybernet's Tactical Ammunition Sorter, developed and deployed for the U.S. Army in Camp Arifjan, Kuwait

---

[19]"Under the Navy and MDA SBIR programs, Silicon Designs Inc. of Issaquah, Washington developed the 'accelerometer' used in most DoD missile systems. . . . Total sales of the accelerometer to DoD and commercial customers are $40 million per year. DoD's initial SBIR investment was just $1.2 million." See Web site cited in footnote 18.

He concluded by asserting that the SBIR program is "very important for the DOD," and for every agency, as "part of an overall collection of things where we try to make sure that we really do have the best technology for the warfighter." He noted that the DoD's SBIR program worked well with both large and small firms, universities, other federal labs, and U.S. coalition partners. This broad reach was essential, he concluded, "because science and technology is critical to our success and to our future."

DISCUSSION

Dr. Gansler pointed out that the examples of large commercialized systems given above were programs initiated by DoD SBIR awards; later, after development, DoD entities also became the purchasers. He said that in other cases, programs that began under SBIR contracts to the DoD later developed products of commercial value that had little to do with the original defense work. He cited the example of Martek Biosciences, now a profitable biotechnology firm that began work with DoD SBIR funding for space programs. Today, Martek sells commercial products for medical purposes that have little to do with the DoD.[20] Other case studies show that projects also may move from one federal agency to another during technology development, according to the changing capabilities of the technology or changing needs of agencies.

A New Role for Small Business Research

A questioner asked Dr. Gansler how S&T needs had changed since 9/11, and whether shifts in the industrial base since the Cold War would affect the distribution of research capabilities. Dr. Gansler agreed that the two issues were related, and that the structure of research in the defense establishment was now "totally different" since 9/11. After the Cold War, consolidation in the defense industry trimmed the number of large defense contractors to six and reduced the participation of small firms. Since 9/11, the DoD has searched more widely for ideas and innovations and sought the R&D participation of more small businesses. He said the DoD was also seeking the participation of more universities in doing research. Specifically, he noted the need to move beyond the large research universities to the smaller campuses, such as Bowie State University in Maryland, a historically black university he had recently visited. "We need to figure out how to get research done at more schools," he said.

Dr. Gansler then introduced the members of the first panel.

---

[20]Martek Biosciences Corp. develops and sells products extracted from microalgae, including formula for children's milk. <*http://www.martekbio.com/home.asp*>.

The SBIR Program:
Different Needs, Common Challenges

Moderator:
Bill Greenwalt
Senate Committee on Armed Services

Mr. Greenwalt congratulated the National Academies for hosting the program and for gathering such a broad range of participants. He introduced the topic of his panel by emphasizing the diversity of opinions about the SBIR program, urging the committee to talk to the widest possible variety of people. There is no single "right answer" to the question of how well the program works or should work, he said, so that this diversity of opinion will be necessary in making improvements.

He described his committee's long interest in the SBIR program. In particular, Senate members wanted to know what kinds of return could be expected on the initial billion-dollar investment, especially in the form of activities stimulated in the entrepreneurial sector of the U.S. economy by defense projects. In focusing on that question, the committee worked closely with the Small Business Committee and tried to help with legislation, focusing primarily on ways to move the best technologies into the hands of warfighters. This challenge is not unique to SBIR, and he suggested that SBIR procedures could provide insights into other kinds of R&D programs seeking to bridge the so-called "Valley of Death" between technological research and commercialization.[21] He said that the Senate Committee on Armed Services fully supported the SBIR program, and looked forward to seeing more companies move into the defense marketplace to become either stand-alone prime contractors or subcontractors to existing prime contractors.

---

[21]"Commercialization" was acknowledged by several participants to include the transition of new technologies either into applications within the DoD and/or into the public marketplace. This process, especially when the DoD is the primary or only customer, is also referred to as "insertion" and "transition."

Michael Caccuitto
Department of Defense

Mr. Caccuitto thanked the organizers and said that he worked for Mr. Frank Ramos, the Director of Small and Disadvantaged Business Utilization, Office of the Secretary of Defense (OSD). This office reports to the Under Secretary of Defense for Acquisition, Technology, and Logistics. Prior to becoming program administrator for SBIR/STTR, Mr. Caccuitto spent five years in the Office of Industrial Policy (formerly the Office of Industrial Affairs).[22]

A Non-linear Technology Commercialization Process

He began with an overview of the commercialization process, which is one of the four main objectives of the SBIR program. This objective had been receiving more attention than any other aspect of the SBIR program since the 1992 reauthorization. He noted that this represents an interest that had steadily increased.

The SBIR is a three-phase program that appears quite linear. That is, a technology is assumed to move more or less sequentially through three phases, from Phase I (feasibility) to Phase II (prototype) to Phase III (commercialization). In fact, he noted, the process is seldom linear in practice, responding to many overlaps and feedback loops and changing direction according to new capabilities and needs.

Funding sources differ for the three phases, with Phase I and II supported by the SBIR set-asides; Phase III must, by definition, be funded by other sources.

As of the FY 2005 budget, 11 federal agencies were participating in the SBIR program, with the budget of the DoD accounting for over 50 percent of all SBIR money. He joined many participants in noting the diversity of SBIR programs, implemented by different agencies in different ways in order to meet agency-specific missions.

Leveraging Small Business Innovation to Benefit the Warfighter

The general objective of the SBIR program in the DoD, he said, was to harness and leverage small business innovation for the benefit of the warfighter and the nation. The program directly supported two current goals: technology dominance and a stronger industrial base. He reminded the audience that a new Under Secretary, Mr. Kenneth Krieg had just been sworn in, with the possibility of fresh input. During testimony before Congress, Mr. Krieg cited the importance of pro-

---

[22]Mr. Caccuitto also thanked the Office of the Assistant Secretary of the Air Force for Acquisition for allowing him to participate in this event. As a reserve officer attached to that office, he added, he was serving his annual tour of duty on the day of the conference.

viding small-businesses access to the industrial base. He also highlighted the importance of acquisitions from small business to enhance the capabilities of the warfighter.

Mr. Caccuitto also cited a comprehensive set of studies by the DoD's Office of Industrial Policy which had dramatically highlighted the importance of small business in providing future warfighter capabilities.[23] Methodologically, these studies were designed to address Joint Staff functional capability concepts: Battlespace Awareness, Command and Control, Force Application, Protection and, most recently, Focused Logistics. The studies divided these warfighting functional architectures into their hundreds of composite capabilities and identified the technologies that enable those capabilities. A survey of the industrial base was then used to determine which firms could best produce those technologies. Early results from the five major functional architectures revealed that, in aggregate, 36 percent of firms that could supply the technologies were small businesses employing fewer than 100 people.[24] This, he said, indicated a need for quick and timely access to small firms of the most appropriate capabilities.

Within DoD were 10 participating SBIR agencies, 6 of which had STTR programs. (See Figure 9.) The three military departments comprised about three-quarters of the SBIR budget. Some agencies had centralized topic generation processes and some had decentralized processes. Some had procurement activities, some did not; these gained importance in Phase III. Some provided "gap funding," to help small firms move from Phase I to Phase II, and some did not. Most offer a Phase II Enhancement program whereby SBIR funds are employed to match external (Phase III) investment to advance technology toward full commercialization. In all, Mr. Caccuitto said he had counted 155 separate functional activities involved in the DoD SBIR program, making it a "highly decentralized and geographically dispersed program."

Imperfect Quantitative Metrics

To date, the department had used "success stories," examples of what the program had contributed to the efforts of warfighters, as a qualitative measurement of success. These stories, he said, were not hard to find. Only two quantitative metrics were available, however—both used to measure business-generated Phase III dollars—and both were imperfect, he said. One was the DoD SBIR/STTR Company Commercialization report (CCR), which requires firms

---

[23]See, for example, Deputy Under Secretary of Defense (Industrial Policy), "Transforming the Defense Industrial Base: A Roadmap," February 2003. This report highlights the importance of small businesses for building future war-fighting capacity. Access at <*http://www.acq.osd.mil/ip/docs/transforming_the_defense_ind_base-full_report_with_appendices.pdf*>.

[24]This is an even more stringent definition of "small business" than used by SBIR, which specifies firms of fewer than 500 employees.

	Army	Navy	AF	DARPA	MDA	OSD	DTRA	SOCOM	CBD	NGA
FY05 SBIR Budget	$233M	$253M	$315M	$67.5M	$126M	$61M	$4.7M	$13M	$5.8M	$700K
STTR Program	$28M	$30.4M	$37.8M	$8.1M	$15.1M	$6M	---	---	---	---
Topic Generation	D	D	D	C	C	C	C	D	D	C
Proposals to Topics	12.6	13.8	13.1	20.1	13.0	13.5	18.6	15.1	13.6	61.0
Gap Funding	Y	Y	Y	Y	Y	N	N	Y	Y	N
Procurement Activity	Y	Y	Y	N	Y	N	N	Y	N	N
# of Activities	32	26	20	9	32	7	6	10	12	1
% of Phase II Enhance	18%	5%	12%	35%	3%	2%	0%	20%	3%	---

FY05 Budget

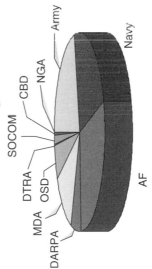

Legend:
- Topic Generation: D = decentralized process; C = centralized process
- Gap Funding: Y = procedures in place to reduce the funding gap between phase I completion and phase II award; N = no gap funding
- Procurement Activity: Y = acquisition program or other procurement activity within Agency; N=no
- # of Activities: number of labs, centers, commands, and/or divisions within Agency that participate in SBIR
- % of Phase II Enhancement: Percent of your Phase II awards that receive Phase II enhancement funding

SBIR in DoD is highly diverse in implementation and decentralized management.

FIGURE 9 Participating DoD components.

that submit bids for Phase I or Phase II SBIR/STTR contracts to report commercialization for all previously awarded Phase II contracts. This collection of commercialization data was becoming, over time, a robust database, measuring both sales and investment derived from extending or logically concluding work conducted during the Phase II and associated Phase I efforts. A principal strength of this dataset is that it contains sales and investment from government and non-government sources, and captures both prime contract and subcontract activity. A principal disadvantage is that data revealing further growth by SBIR/STTR "graduates" (i.e., former SBIR award-winning firms that have become ineligible either by organic growth or acquisition) is not generally captured because these firms are not competing for new awards. The second data source was the DD 350 Form and accompanying database. This is a rather elaborate record of individual contract actions that is required for all direct contracts with the government. A field in this form is dedicated to SBIR, and it is possible to indicate Phase I, Phase II, or Phase III, though it is questionable, he said, based on input from veteran contracting officials within the Department, whether contracting officers in the field pay attention to this discriminator or are necessarily aware when a contract action qualifies as an SBIR/STTR Phase III award. There is currently no way to capture Phase III activity via subcontract reporting mechanisms. In summary, he said, measuring commercialization remains a considerable challenge.

He discussed the progress of commercialization of SBIR firms over time, based on the DoD SBIR/STTR commercialization database for firms participating between 2000 and the present. A chart (Figure 10) showed that revenues

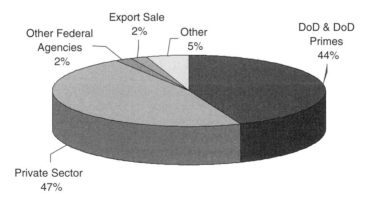

Non-defense sales and investment slightly exceed that of DoD and DoD primes combined—significant spin-off achieved by balanced investment program.

FIGURE 10 SBIR Phase III commercialization.
NOTE: Reported by firms submitting to DoD in 2000-2005, DoD projects only.

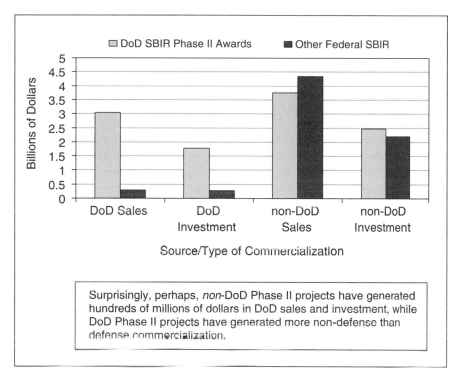

Surprisingly, perhaps, *non*-DoD Phase II projects have generated hundreds of millions of dollars in DoD sales and investment, while DoD Phase II projects have generated more non-defense than defense commercialization.

FIGURE 11 Phase III composition: Total sales and investment of DoD Phase II Awards
SOURCE: Department of Defense CCR, 2005.
NOTE: Sixty three percent of projects are DoD Phase II awards, and 37 percent are other Federal Phase II awards, indicating a high degree of capture of non-DoD Phase II commercialization data.

generated by commercialization had, over a 20-year period, outpaced the size of the SBIR budget. He noted a four-year lag, but said this could be expected because of the time required to commercialize. Another chart (Figure 11), plotting sales against investment, showed that from about eight years after a Phase II award, sales begin to outpace investment. This graph was "a bit cautionary," he said, showing the need for patience in developing and realizing commercialization.

Mr. Caccuitto went on to show that DoD was also investing in and buying the products of non-DoD Phase II projects. The dataset showed that 37 percent of the DoD's Phase II commercialization records were generated by contracts awarded outside the DoD. These observations together meant, he said, that "we're leveraging work that's being done by other agencies. And we're beginning to capture in this database a significant percentage of the non-DoD SBIR projects."

The Advantage of Requirements "Pull"

He said that he did not want to sound simplistic in confusing technology "push" versus mission "pull," but he had found, from working in both a program office and in a laboratory environment, that it is more difficult to push technology than to pull it, even with a "good requirement." He used an example from when he was an advanced technology project manager in the Air Force's E-3 Airborne Warning and Control Systems (AWACS) system program office. He described the competition for resources that takes place in a program office environment, where advanced technologies are weighed against other system needs when determining how limited funds are spent. This competition for resources is driven by warfighter priorities, which may not favor the capability provided by the advanced technology over other needs. In other words, a requirement alone does not guarantee funding. Bridging that gap can be difficult because of this resource limitation, and also because R&D, by SBIR definition, does not explicitly include T&E (test and evaluation). This means that the SBIR program tends to only develop a technology to a readiness level of about four or perhaps five, whereas a program office may not be willing to consider it until level seven or eight. This effect is further explained by observing that later-stage technology maturation (T&E) tends to be quite resource intensive. Thus, tapping the limited pool of funds set-aside for SBIR/STTR alone may not be the most effective way to address this issue. Indeed, bridging this gap was the challenge, he said, that the symposium was trying to address.[25]

Technology Insertion as a Team Effort

Mr. Caccuitto turned to a broader look at the overall S&T portion of the defense budget, which he called "bleak." Looking ahead, he said, he saw only increasing competition for defense budget dollars, which would make his job and the job of technology development and transition generally more challenging. Based on his experiences, he suggested that a successful technology is the product of a complicated equation involving many different stakeholders in the acquisition community, the government S&T community and large and small businesses. "Each has a part to play," he said, "and it takes champions in each of those places to actually make it work."

He described measures the DoD had taken to improve the program, highlighting the fast-track program, which was intended to speed up Phase II for companies that could demonstrate matching dollars from a third party. He also pointed

---

[25]The analogous challenge in a civilian environment would be to develop the commercial value of a technology to the point that it is ready for commercial sale. Unlike development for a government customer, this requires a host of business skills, such as marketing.

to the sponsorship of topics by acquisition programs, initiated in 1999. Currently, DoD had a majority of topics with sponsorship or endorsement from acquisition programs. His office established a commercialization achievement index in 2000 to measure how successful the proposing firms were at commercializing their SBIR investments, and has been using this information in their source selection process.

Among his office's current efforts was a study being conducted with the RAND Corporation on Phase III, focused on Major Defense Acquisition Programs (MDAPs). The study used a comprehensive questionnaire and follow-up interviews with the Department's MDAPs. Its purpose was to characterize and benchmark best practices relative to technology transition in the SBIR program, with an eye to improving the program and informing policy development.

Mr. Caccuitto concluded by saying that his office and the National Research Council were beginning follow-up review of the fast-track program. Several years earlier, the DoD SBIR/STTR program received a fairly low grade from the Office of Management and Budget's Program Assessment Rating Tool (PART). This assessment focused on commercialization, and the program had taken some steps that had been recommended to modify commercialization achievement index utilization and improve capture of commercialization data for program administration.

Michael McGrath
U.S. Navy

Dr. McGrath said that his job as Deputy Assistant Secretary of the Navy was to help acquire technologies developed by small firms and to help those firms advance their technologies across the Valley of Death. Thus, he had a strong interest in Phase III and saw, like other panelists, considerable variations in the SBIR program even within the DoD, despite a general set of overarching rules shared by all programs.

SBIR as a Source of Innovation

He began with the point that the Navy views SBIR as an important source of innovation across the whole RDT&E, or research, development, testing and evaluation spectrum. Second, the Navy emphasized the "D" more strongly than the "R," like most agencies, and accordingly his office had implemented a process to aid the transition of SBIR technologies into full application. This development emphasis led to SBIR products that were well positioned to respond to "pull" from the program office. A third point was that SBIR gave him an unusual degree of execution-year flexibility. For most RDT&E accounts, the Department had to include in the President's budget a description of how the money would be spent, with little deviation possible. With SBIR, they had flexibility right up to the year

Percent	BA	Title
3	6.1	Basic Research
6	6.2	Applied Research
7	6.3	Advanced Tech. Development
21	6.4	Adv. Component Dev. Prototypes
51	6.5	System Dev. and Demonstration
2	6.6	RDT&E Management Support
10	6.7	Operational System Development

16% → (3, 6, 7)
84% → (21, 51, 2, 10)

FIGURE 12 Where do the SBIR funds come from? Across the RDT&E spectrum.
SOURCE: Data from Navy FY03 SBIR Assessment.

of execution to select topics and emphasis. In addition, most R&D programs in the Department had to be planned years in advance in the budget cycle. The SBIR projects did not, and as a result they brought the critical ability to move technologies more quickly to transition. The Navy has found that its average cycle time, from nominating a topic to getting a Phase I award, was about 14 months. The Navy was working with Dr. Holland's Office of Science and Technology to shorten that time.

Dr. McGrath noted that most SBIR funds came from the 2.5 percent assessment on extramural RDT&E accounts, supplemented by funds taken from the general RDT&E account for program management and execution. (See Figure 12.) In the DoD, research and development activities are classified in 6.1, 6.2, and 6.3 accounts, as described by Dr. Holland, with the most "basic" research funded in 6.1, and the most "developmental" work in 6.3. These accounts paid for 16 percent of SBIR program money, with the other 84 percent coming out of accounts that fund testing and acquisitions (6.4 and 6.5) and upgrade of fielded systems (6.7). He repeated an earlier observation that research and development was not a linear process, but that the classification process was approximate and left many opportunities for innovation and revision at all stages.

An Effective Technology Assistance Program

As a routine, the Navy participated in every DoD SBIR topic solicitation—three per year for SBIR and one per year for STTR. Some of the topics included in those solicitations came from the S&T community, but 84 percent come from the acquisition community, systems commands, and program executive officers.

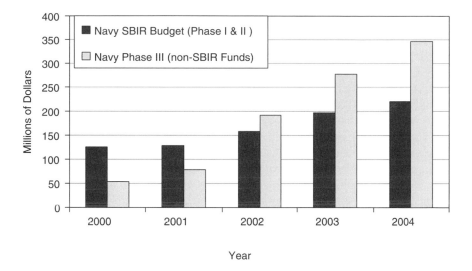

FIGURE 13 Navy Phase III success is growing.
NOTE: Phase III funding is from OSD DD 350 reports, and may be under-reported. FY04 Navy Phase III ($346 M) comprises 114 separate contracts to 81 individual firms.

The program executive officers in the systems commands participated in selecting the proposals and managing them through Phase I and Phase II, always with a view toward the Phase III transition. By this method of asking systems commands for topics, a "pull" had been created from the programs in a position to fund Phase III. He had concluded that small businesses often have a difficult time dealing with DoD and with prime contractors in transitioning the SBIR results. To ease this problem the Navy had implemented a program called the Technology Assistance Program (TAP); every Phase II SBIR contractor had the opportunity to participate in that, and most of them did. The Navy also holds an annual opportunity forum, which in May 2005 was attended by nearly a thousand people and about 150 companies. Attendees included not only PEOs (Program Executive Offices) and program offices, but also prime contractors and outside investors looking for commercializable results.

He showed a chart showing the growth of Phase III funding (Figure 13). The chart was based on DD 350 data beginning in 2000. Although these data did not include contracts issued by prime contractors to small firms, but they did use a form that allowed contracting officers to report government contract awards. The chart shows a persistent and rapid growth of Phase III contract amounts for five years, outpacing Phase I and II awards. He said that because these Phase IIIs are probably under-reported in the Navy, the trend result is probably conservative.

Role of the Prime Contractors

He then discussed the prime contractors, emphasizing their essential role in the SBIR program. In most cases, he said, a small firm would not be able to win a Navy contract unless a prime contractor was involved at some point. About five years ago, the Navy had started a "Primes Initiative," an active outreach effort to connect program executive officers with prime contractors in a more formal way. They had found that the prime contractors see great value in partnering with SBIR companies, partly because the prime contractors want to leverage the work of small firms, influence SBIR topics, and join the SBIR in other small and disadvantaged business programs. The prime contractors also wanted better insight into SBIR data, including places where awards are made and information about the SBIR Phase I and Phase II projects. He noted that the prime contractors deal with all the services, and would benefit from a pattern of uniform procedures across the services. This will require closer consultation among the services about ways to standardize their requirements and systems.

He summarized by saying that SBIR provides a unique source of execution-year funding that was proving adaptable to extraordinary demands. For example, during the previous year the Program Executive Offices had made a concerted effort to respond to the urgent problem of Improvised Explosive Devices (IEDs) in Iraq by creating a "quick response topic." In five months they had managed to make 38 Phase I awards, and 18 of the firms had already progressed to Phase II proposals. Some of those had been awarded, and Phase II prototypes would be tested in theatre over the next six months. He offered that example of using SBIR "in a somewhat unusual fashion" to obtain quick results in the form of prototypes in the field. "To the extent that those prototypes meet the military needs," he said, "I would expect a number of them to move on to Phase III."

He concluded by saying that from the Navy's standpoint, "SBIR is all about Phase III."

Mark D. Stephen
U.S. Air Force

Mark Stephen, Chief of the Air Force's Science and Technology Division at the time of the conference,* said he would not talk about Air Force priorities or show statistics, many of which had already been reviewed, or even relate success stories, though there were many. Instead he said he would address the issues in the Air Force SBIR program that could be improved to make the program more effective. He began by saying that from a mission perspective, the purposes of the SBIR program were (1) to stimulate innovation today that would help the warfighter of tomorrow, and (2) to make sure the program and the acquisition

---

*Now with Coleman Aerospace.

process are optimally effective. Ideally, he said, the SBIR should bring in new ideas more rapidly than the traditional acquisition process.

The topics generated by the Air Force had mainly concerned technology areas and lower-level research. Unlike the Navy, he said, which had focused more on "the D" (development) that more directly expedites the transition process, the Air Force had focused more on the lower-level, 6.2-level research areas.

SBIR Programs Do Not Cover Administration Costs

He raised the issue of the cost of administration. The SBIR legislation does not permit the use of SBIR funds for running the program itself; it must all go into SBIR contracts. Administration costs have to be paid directly by the services, which was one reason the Air Force had been reluctant to invest in staff.

He said that this rule had hampered the effort of the Air Force SBIR program. Their entire program was being administered by only four people at Wright-Patterson Air Force Base, in Dayton, Ohio, who supported nine technology directorates, the product centers, the test centers, and the logistics centers that buy the technology as it matures to the point of program insertion. The great majority of the topics were written by the technology directorates of the Air Force Research Laboratory.

The Air Force SBIR had grown 70 percent in the previous five years without investing in improving the transition process. Because this is what small businesses need, it is also where additional program manager effort must go. One approach, he said, is to ensure the interaction of the program officers and the prime contractors. The office was developing a training and education program for them now, designating people to do that and generally to focus on buying the Phase II projects and aid in the transition to Phase III.

A Need for More Staffing

Since the Air Force program was so thinly staffed, he said, there was a need to improve execution—raising the obligation rates in the program by moving the money to the small businesses more rapidly. There was also a need to cover the costs of running the program so that they could focus more on transition. To do that, he said, his office needed to ensure that there were topics focused on the programs of record and executed by the program officers in the product centers. The program officers were the people within the Air Force who would actually buy the technology, so the SBIR program had to be sure SBIR activities would interest them and that they knew about SBIR. Educating those program officers would require more manpower and training, as the Navy had shown. Program officers who knew more about the SBIR program could then find technology that was important to them, especially among Phase II programs that were starting to mature, and communicate the advantages of the SBIR program to other program managers and program executive officers.

Improving Topic Generation

Col. Stephen said his office could improve topic generation by initiating planning farther in advance, but this would decrease the timeliness of topics and reduce the chance that technologies would be state-of-the-art when ready for insertion. His office, after a study of execution, had discovered that one of the greatest needs was to shorten the topic generation time. They, like the Navy, also found the advantage of awarding amounts to some Phase II projects larger than the traditional $750,000. That figure, he said, was not always enough to develop a technology far enough "to excite the program officers," and the Air Force would be offering larger amounts.

He also said the Air Force needed to re-organize and restructure its SBIR office by adding staff, increasing effectiveness and efficiency, and increasing its obligation rates. They were adding emphasis on the product centers and the prime contractor involvement, which was needed to ensure transition within Phase III. He said that it was the responsibility of his office, rather than the laboratories or product centers, to improve the transition process.

He also addressed the issue of success stories, a qualitative measure of program effectiveness. The Air Force program had no funds to document or track its successes. Some stories were relayed to them by participants—in some cases by contracting officers who checked a box on the DD 350 form—but to date these stories had been "captured mostly by coincidence." One recent success story had been a project on which the Air Force worked quickly with the Army and the Navy to produce a new kind of body armor and move it quickly to the field to support the warfighter.

He said that the Air Force needed and wanted to improve its SBIR program, and asked for help from the participants. He concluded that the ideas described in the conference would all be useful to educate the product centers, generate better topics, and sharpen the selection process.

John A. Parmentola
U.S. Army

Dr. Parmentola began with a brief sketch of his background. At one time, he performed basic research at a university, building a research program at an institution that did not have one. He then moved to teaching science and technology policy at the John F. Kennedy School of Government at Harvard University, and later worked for a private sector corporation as an inventor. He said that he has developed technologies all the way from basic concept to application in the field. "That in itself is an experience," he said, "in terms of understanding all the difficulties it involves." He had spent the last several years as Director for Research and Laboratory Management, managing all of the Army's basic research and about half of its applied research. Trying to promote change from his current position within government, he said, was one of his greatest challenges—especially since he was also a small business owner who understood the problems faced by small businesses in gaining a foothold with government customers.

As Director for Research, he reported to the Chief Scientist for the Army, who worked for the Assistant Secretary of the Army for Acquisition, Logistics, and Technology. He worked closely with the Director of Technology and a Director for S&T Integration. He was responsible for basic and applied research, as well as for the laboratory infrastructure over several major commands: Research, Development and Engineering, of which the largest was the Army Corps of Engineers; Army Research Institute; Medical Research and Materiel Command; and Space and Missile Defense Command. The small business program manager worked within the Army Research Office, located in Washington, D.C.

He said that his diverse background placed him in a good position to help all parties with the problem of transition. The Army had taken several experimental steps that "may have some promise for moving the small business innovation and transitioning."

In terms of process, he said that successful Phase II projects could pass directly into the military as components for Army technology objectives, which themselves are under the control of project managers, or they might go directly to Program Executive Offices or to other services' depots. Others would move to Phase III for additional development. He mentioned examples where SBIR projects have actually gone to support a need at a military depot. The projects might be paired off with prime contractors, or they might be commercialized directly, preferably offering some benefit to the military.

The Function of OnPoint

He described a recent Army venture capital initiative called OnPoint, whose mission was to invest in small, entrepreneurial companies, including those that would otherwise not do business with the Army. OnPoint, which focused on mobile power and energy for the soldier, had the dual objectives of (1) helping the transition of technology and (2) earning a return on investment. In effect, it served to triangulate between small companies and potential customers, which were usually prime contractors that already had a contract with a program manager to supply a product. Dr. Parmentola's office sought to take successful Phase II projects and provide OnPoint with descriptions of those concerning power and energy. The projects are placed in a pool, along with the other proposals, from which OnPoint selects some for funding. The Army does not influence this process other than to supply OnPoint with information.

Another important role of OnPoint is to manage the often conflicting expectations of the small firm within the confusion of the acquisition process. The small entrepreneurial company usually seeks either to provide a component technology or move directly and quickly into an acquisition program. He characterized small businesses as speedy Ferraris which expect to move fast and earn revenue, but which often find themselves blocked on a narrow road by a hay truck—the acquisition mechanism. OnPoint attempts to foresee and resolve such blockages and conflicts.

Barriers to Success

He said that the system does have inherent barriers that threaten the success of Phase III. First is the inherent risk in dealing with programs that are new and have not yet done business with government. There is a steep learning curve for these firms, some of which may be as small as two or three partners working on an idea.

A second barrier is that small firms seldom have sufficient resources in people or money to do the market analysis needed to catch the interest of a venture capital firm. OnPoint tries to help with this, although the process was still experimental and still faced challenges of planning, programming and budgeting. Another challenge is that the DoD planning, programming, and budgeting system works on a two-year cycle, and it is difficult for program managers to calculate whether a small entrepreneurial start-up company will be able to produce a product fast enough to fulfill a need.

The program also faces funding constraints, with a maximum of $850,000 allowed for SBIR Phase II awards. This may not be enough to produce a prototype of sufficient maturity to move into a program manager's ongoing program. Program managers are notorious for avoiding risk, he said, and if a technology is not ready, they are reluctant to accept it.

Finally, he noted a paradox in that the system needs both rapid commercialization and long-term research to provide the revolutionary but unforeseeable capabilities that fulfill future needs. The department must try to strike a balance between these conflicting strategies, which it attempts to implement through the pilot program with the Program Executive Offices and through maintaining an emphasis on long-term innovation in DoD laboratories.

In summary, he said that the Army was providing multiple paths to increase the chances that innovations produced by small businesses will be able to move into military systems in support of U.S. soldiers.

Carl G. Ray
National Aeronautics and Space Administration

Mr. Ray observed that, based on the evidence of the talks so far, all the agencies faced some of the same challenges. He said he would not repeat these challenges, but try to present a more abstract description of the NASA program.

He said he would give a perspective on what NASA views as the ultimate outcome of the SBIR process, vis-à-vis the intent of the SBIR legislation. That is, if the agency can use government funding to produce good technologies of value to agencies, many of those technologies can then move toward the commercial marketplace where they grow into something larger and contribute to the national good.

He also said that from NASA's strategic perspective, it is difficult to move a small firm to Phase III from the inside of government. Even though the agency provides awards and tries to facilitate the direction and activities of a firm's development, these actions tend to remain heavily focused on agency, rather then firm, needs. As new products move toward technological maturity, from discovery through development and commercialization, the process is not linear, even though it may be depicted in that way. In reality, it is complicated by many gaps and challenges between stages.

He also described the impacts of an innovation in terms of its position in a "technology life cycle." (See Figure 14.) He noted that an innovation can occur at any point on that life cycle. The point where an innovation occurs creates a "disruptive" technology, where the innovation can essentially create a development path of its own. Farther along the path toward application, one finds that new pathways may be created when new products spin off from the basic technology under development. Innovation may come at an even later stage, constituting an extension of the technology to produce a base technology in the marketplace. Even though these dynamics are difficult to manage or even understand, he said, they are useful to be aware of because they showed why a product's utility is related to its technological maturity.

Also, he said, the SBIR can be mapped by using a business approach. Where technical points of innovation occur, business opportunities may also occur, depending on when that innovation occurs. From a business or financial standpoint, one may look at partnerships or the SBIR program itself as an opportunity. Farther along, one may consider licensing and then patents, which may assist in moving a technology product forward. The challenge, he said, is to manage all of these technical, financial, and commercial realms.

The Three Phases of the SBIR Program

In terms of the program itself, the SBIR program at NASA, as in other agencies, is a three-phase program. (See Figure 15.) NASA begins by aligning the SBIR program with the needs of its own mission directorates. The topics and subtopics that flow from the directorates to the SBIR program start with a basis of traditional NASA strategic planning and annual mission planning. The needs of the missions are then described in the topic solicitation, which is smoothed and expanded so they can more easily be understood by small business applicants.

Once a solicitation takes place and the small businesses move through Phases I and II, the agency starts, through contracts and other mechanisms, trying to educate the companies to think about not only the technology itself but also about

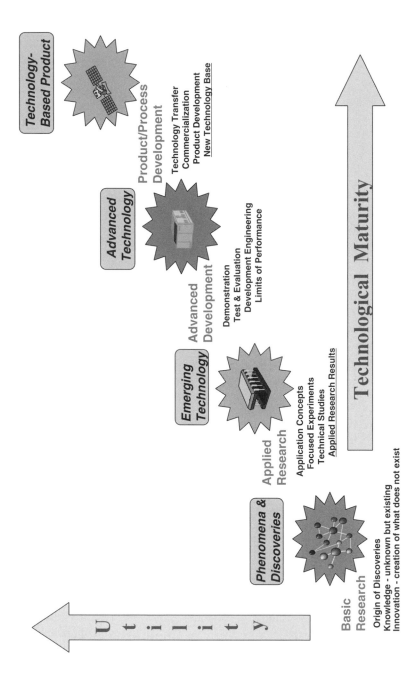

FIGURE 14 Paths for new technology products.

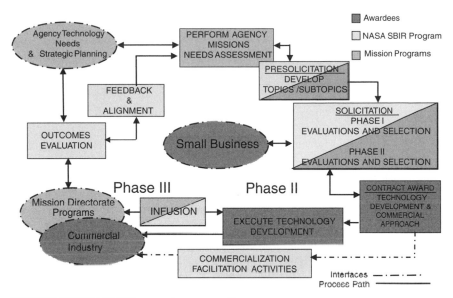

FIGURE 15 SBIR/STTR strategic program flow.

the commercialization aspects. This process begins by examining sub-topics for the best possible quality—looking internally at the customer base, which, for NASA, is its four mission directorates: the exploration systems, space operations, aeronautics, and pure science.

As Phase III begins the agency tries to facilitate a firm's movement toward the commercial arena. This consists of helping companies find ways to advance themselves in the marketplace by creating assistant programs or finding organizations and assistant areas that already exist.

One of the differences between NASA and DoD is that DoD, in Phase III, might buy hundreds or even thousands of a product from a small firm. Because NASA does not build fleets of Space Shuttles, say, or other complex products, it cannot sustain a small business with such bulk purchases. So NASA must use other measures for partnering with organizations that assist small businesses, such as the SBIR/STTR 2005 National Conferences that link SBIR firms to business opportunities and the investment community.

Helping Small Businesses Understand Agency Needs

Some NASA initiatives are designed to help companies understand agency needs and NASA nomenclature, which differs from that commonly used in the marketplace. The agency has built a "taxonomy table" to translate "NASA-ese" to more widely known terminology. It has also produced "hallmarks of success" videos and success stories by visiting SBIR companies on a regular basis. Initiated about two years ago, this program invites companies to discuss their relationship with NASA and the outcomes of their R&D efforts. This lets the agency learn what can be done better and gives the companies an opportunity to advise other companies on how to be successful in the program.[26]

NASA also holds national conferences, in collaboration with the other agencies, that have drawn interest from the investment community, and they have created a NASA Alliance for Small Business Opportunity (NASBO). Although NASA does not have a "fast-track" program like that of DoD, NASBO is a new program that seeks to create some fast-track elements. Its main objective is to find investment facilities or other capabilities and form partnerships with them. These might be a state entity, an angel investor, or a venture capital firm.

In addition to traditional commercialization, NASA tries to find other opportunities to leverage their companies' products by helping form partnerships with the mission directorates. The agency calls this process infusion, or "spin-in": using the internal mechanisms of the mission programs to help move technology into the mission directorates. To expedite this process, the agency uses a variety of virtual program aids, including tools for project presentation, demonstration development, risk analysis, and infusion planning. The program is able to achieve closer communication and more support from the directorates, especially in finding the right kind of sub-topics. It is still working to educate both the companies and the mission directorates about how to move SBIR-developed technologies and their companies into a position where the directorates can utilize them.

A difficulty in the infusion concept is that program managers still have to be convinced that an SBIR technology can be integrated into their programs, both programmatically and technically. They have to be shown that this can be done in a timely fashion, and in a way that will minimize the risk of using that technology. They must also believe that the SBIR firm understands the directorate's technical challenge and that this challenge will not reduce the performance of the technology.

As illustrations of success, Mr. Ray showed slides of several SBIR technologies adopted for use on the Mars Exploration Rovers, including an ASCII chip from Maxwell Technologies, of San Diego, California; lithium ion batteries from Yardney Technical Products, of Pawtucket, Connecticut; and heat switches from Starsys Research, of Boulder, Colorado. He closed by saying that a NASA commercial metrics survey published in 2002, covering the program period from 1983 to 1996 and 75 to 80 percent of the NASA SBIR firms, showed that a minimum of 38 percent of the NASA Phase II awards demonstrated strong commercial intent in nongovernment markets.

---

[26]See <*http://sbir.gsfc.nasa.gov/SBIR/success.htm*>.

DISCUSSION

Mr. Greenwalt offered a summary of the main point discussed thus far—that a small SBIR business, after completing award-funded Phase I and Phase II work, faced a large financial barrier before it could accomplish the developmental work necessary to complete Phase III, or commercialization, successfully. For a company doing long-term research, gaining access to venture capital could require a wait of a year or a year and a half. A company ready to do rapid innovation would seldom find Phase III money without the participation of a prime contractor. There seemed to be no specific mechanism within or related to the SBIR program to help with this transition.

Few Small Businesses Cross Phase III

He said he was forced to conclude that even though Phase I and Phase II of the SBIR program are generally successful, and some firms have been able to cross the Phase III barrier, the vast majority of SBIR programs have not. He reported that Congress receives many inquiries from frustrated small businesses who say, "'I have a great idea, and I won a Phase I award and a Phase II award, and everyone agrees that my technology is a great thing, but there's no support for me.' And they fall into the Valley of Death."

He said he was left with the question, What can be done about that? Is the SBIR program funding too many small firms, or is there a need for transition funding at the agencies to take better advantage of the products of an otherwise sound program?

He asked the panelists and others to respond to these questions: What can be done to help these companies, and to help the country gain access in the best way to new technologies?

Options for Small Business

Dr. McGrath, referring to the earlier characterization of the acquisitions process, said that a small business has two options: drawing on Dr. Parmentola image of the small business Ferrari and the agency procurement hay wagon, he noted that one option is to get around the "hay wagon" blocking the road and move ahead quickly. The other, more common option is to jump aboard the hay wagon, which is the program of record. He suggested that from the small business standpoint, some improvements might remove the barriers and allow small firms to move quickly to real commercialization, which would be a good thing. But a more feasible goal for the agencies would be to help firms join existing programs.

He recalled earlier discussions of technology readiness, and the fact that federal programs tend to be risk-averse. He advocated more latitude in the interpretation of RDT&E and being able to qualify items for use. An important part of a solution would be to qualify SBIR activities as part of development, thereby quali-fying them for some amount of developmental testing. In placing a boundary on legitimate targets for SBIR, he proposed a broader definition of technological readiness.

Col. Stephen agreed, and said he had seen the same problem in the Air Force program. He suggested that rather than providing a separate pool of funds for Phase III, the program should (1) increase the amount of the Phase II award, (2) increase the emphasis on development, (3) educate the program offices to make sure they are aware that a particular technology is being developed, and (4) teach program offices to view the SBIR program as an integral part of their research and development effort, not something "extra."

Incentives for Program Managers

Mr. Greenwalt asked what type of incentives might be used for program executive officers and program managers. He recalled that six or seven years earlier he was hearing the complaint from program management that the SBIR program was essentially a program "tax" that brought them no benefit. He asked what type of incentives would generate buy-in from these program managers.

Col. Stephen suggested that in order to gain buy-in, the program should be sure to focus not only on research, but also on the results that program managers need—outputs that directly support agency objectives.

Dr. Parmentola agreed, saying that program managers want technologies that have been operationally tested and require little, if any, modification. The reluc-tance to work with the SBIR program, he said, is the same reluctance expressed about S&T in general: Managers are concerned primarily with technology readi-ness levels, not with long-term research. They have schedule and cost constraints and have to be convinced that these will not be disrupted if they adopt a technol-ogy that is not proven. They also have to believe that they will be able to coordi-nate the S&T with manufacturing the product in the quantity needed, with reli-ability and quality control.

Research is Seldom Efficient

He said that everybody wants greater efficiency, but he had not convinced himself that the R&D process could be made much more efficient. His own back-ground was in the basic research community, where he had experienced a high degree of inefficiency. He explained that a typical researcher can imagine more ways that the world doesn't work than ways it actually does work. So for a typical research program, the odds are inherently against finding a valuable new product or process. Anyone who has tried to take a concept from "cradle to grave," from discovery to commercialization, he said, understands how many things can derail a development project and even cause it to fail.

He said he had not learned how to make R&D much more efficient, but that it was probably possible to provide some help and some new pathways to increase the likelihood that a technology will add to a transition. One way is to enter a small firm into S&T programs that a program manager is already engaged in. This can ensure that the laboratory understands what the program manager wants, can work closely with the program manager and have a better idea of how to package its technology.

Mr. Ramos referred to Mr. Caccuitto's discussion, saying he was studying such systems as the joint strike fighter, the Virginia-class submarines, and future combat systems in search of ideas that could improve and extend the life cycle of the platform. He said that Mr. McNamara of the Navy had an excellent example of inserting technology for SBIR into Virginia-class submarines, using some incentives. He urged the group to think about Mr. Caccuitto's discussion of technology dominance and supplier base, and find ways for small businesses to fit into that stream. He predicted opportunities for some "ad hoc R&D" if people took a broader view of where to make this transition and better manage topic selection.

Dr. Wessner said he had heard five points that might be useful to list for comment:

1. What is the prevailing attitude of program managers towards the SBIR program—is it an opportunity or is it a tax?
2. Regarding the alignment issue Dr. Parmentola raised, how can R&D in the SBIR program be better aligned with the expectations of program managers?
3. How can we raise additional funds for testing?
4. If the award amounts are too low, as several participants had said, should there be [a] fewer awards, [b] larger awards, or [c] an increased set aside?
5. There is presently no funding for SBIR management—how should that be handled?

Dr. McGrath commented that the SBIR program was indeed viewed negatively as a tax in the Navy, and probably elsewhere in DoD. He said that John Williams of the Navy had played a large role in changing that attitude over the past six years or so by delivering the following message to the Navy's program offices: "The SBIR program provides money and opportunity to fill R&D gaps in your program. Apply that money and innovation to your most urgent needs." He said that this approach had placed the SBIR program in a different light for many people.

He also commented on the last question, regarding management. The lack of funding for management was a limiting factor in the Navy because most of the systems command effort is funded on a reimbursable basis. It requires managers to take dollars out of other programs to pay the people who manage the SBIR program, which is "painful."

Changing Negative Attitudes about the Set-aside

Col. Stephen said that Air Force program officers had also felt that SBIR money was money "taken out of their hides." The normal routine is for the program office to submit a budget that is based on an independent cost estimate of program needs. They must then take 2.5 percent out of this budget with the mandate of doing something else with it. He said that a better approach would be to make the program officers feel that the set-aside *is* their money, and is part of their R&D program. They can use the money to write topics they are interested in. Rather than increasing the whole pot of funding, he suggested raising the grant limits. This might reduce the number of SBIR awards, but it would make each project more viable and give more value to the program officers, who could then expect a higher level of technology readiness for their program. They could even use some of those dollars—*their* dollars—to hire people to help manage the SBIR projects and to advocate for the SBIR program within the program office. Those steps could be taken, he said, without changing the current set-aside rate of 2.5 percent.

Streamlining the Acquisition Process

Mr. Carroll returned to the issue of the "hay wagon," the complex acquisitions process. Fitting SBIR into that acquisition framework, he said, is very complicated, but he said he thought it could be done in a way that might even streamline the process itself. He referred to a proposal to Congress by Dr. Gansler six or seven years previously, which recommended that at least half of SBIR projects should be tied directly to acquisition programs. He said that this goal made sense, along with educating the program management personnel on how to use and benefit from SBIR. This had been part of the training curriculum for a year, but had been deleted. He suggested re-examining the objectives recommended in that earlier proposal which "still make a lot of sense." He said that the Navy had followed a number of those recommendations with good results.

Mr. Caccuitto responded that he had just reviewed the SBIR numbers for 2002 through the end of 2004 and found that over 65 percent of the topics had endorsement or sponsorship from acquisition activities. He said that this was a result of Dr. Gansler's directives in the late 1990s. Even with that result, he said, it was still difficult to integrate SBIR into acquisitions activities, which underlined the importance of taking steps to enhance the process.

Transitioning SBIR:
What Are the Issues for Prime Contractors?

Dr. Gansler, standing in temporarily for the moderator, observed that while the discussion today focused primarily on DoD and NASA—agencies that buy extensively from the SBIR companies they support—the overall National Academies study would also address the programs of the other agencies, which purchase less from their former grantees. For those agencies, he said, the committee would still have to address the question of how to better align the SBIR programs with agency activities.

Moderator:
Max V. Kidalov
Senate Committee on Small Business and Entrepreneurship

Mr. Kidalov began by seconding the remarks of his Congressional predecessor, Mr. Greenwalt, to reaffirm the interest of Congress in the success of the SBIR program in general and in Phase III commercialization in particular. He introduced the panel members and suggested several themes for the discussion. He said Congress was interested in hearing about (1) the challenges and obstacles to integrating SBIR firms and SBIR products into systems and platforms developed by companies, and (2) best practices that had been observed by the panelists. He expressed interest in hearing about minority assistance programs and programs such as the mentor-protégé program, along with any lessons the SBIR program could learn from them. Finally, he encouraged more discussion of intellectual property and its role in the relationship between prime contractors and subcontractors.

Richard H. Hendel
Boeing Corporation

Mr. Hendel began with a description of Boeing Corporation. Its largest segment is the Commercial Airplanes Organization, and next in order is the Integrated Defense Systems, created several years ago out of two divisions, Military Aircraft and Missiles, and Space and Communications. The company also has a "Phantom Works" group that performs a large part of the company's research and development and that initiates engineering technology efforts and new programs. The company manufactures many defense-oriented systems, including military aircraft, transport aircraft, bombers, weapons, space and communications, large-scale, integrated future combat systems, and advanced technology projects. The company functions in many locations in the United States and abroad; the Phantom Works organization is headquartered in St. Louis with personnel in four or five other locations. Phantom Works develops such projects as advanced systems, prototyping, the unmanned combat aerial vehicle, and many exploratory concepts.

Long Involvement with SBIR

Boeing's involvement in SBIR dates from the years 1991-1992, before the merger between Boeing and McDonnell-Douglas; since then the joint company's SBIR activities have been merged as well. Most of Boeing's interaction with small firms through SBIR has occurred in the Phantom Works (PW) program, which he said had done a good job of supporting them. Boeing personnel were currently working with small businesses on 27 SBIR contracts: 4 in Phase I, 22 in Phase II, and 1 in Phase III. He estimated that over the years, Boeing had worked with nearly 200 SBIR projects. Their interaction included support in the form of follow-on with the companies and tracking the development of their technology. He also participated in national SBIR conferences, such as the recent Navy Opportunity Forum in Reston, Virginia.

He said that Boeing's management had recently decided to increase the emphasis on SBIR. One result of this increase in emphasis, he reported, is that he had been asked two months earlier to increase the time he spends on the program from 25 to 100 percent.

Boeing's SBIR Procedures

Boeing had developed its own SBIR procedures. One was to poll all of their technologists and researchers to review the SBIR topics at primarily four agencies—DoD, NASA, Homeland Security, and the National Science Foundation—and report on any that interested them. The office would then assemble a list of those projects and share the list with small businesses, both at conferences and by an external mailing list. The intent is to provide a point of contact at Boeing where small businesses can talk about their interests and those of Boeing, and how they might fit together. The list also helps Boeing track the technologies being developed by the companies. If asked, the company will provide a letter of interest and support for their Phase I and Phase II proposals. Sometimes the companies ask Boeing to collaborate with them, such as when they lack certain facilities or other capabilities. Companies have used the flight simulation labs, for example, and asked for other assistance via a statement of work. Such relationships are advantageous to both parties because Boeing is a potential customer for the technology being developed.

The office was trying to track all such involvement, along with the results, and issue to Boeing management a quarterly status or activities report on all interactions with small businesses, including any efforts to advance the program internally. While awareness of the SBIR program was high in the Phantom Works, Mr. Hendel wanted to expand this across the Integrated Defense Systems and its large programs, such as the F/A-18, the Joint Direct Attack Munitions, the C-17s, and the Delta launch vehicles programs. There he had found limited awareness of the SBIR program, and he wants to elicit more involvement from those programs.

He mentioned that over the years, Boeing had been involved in submitting topics to the agencies, some of which end up in agency solicitations. The best way to meet technology needs, he said, was to develop more collaboration between the programs, and between the small businesses and the large businesses. He said that both the Multi-mission Maritime Aircraft and Future Combat Systems programs were very interested in having Boeing submit potential topics for them to evaluate as candidates for solicitations.

Some Boeing Success Stories

He offered some recent Boeing success stories, which reinforced the points that SBIR is not a linear process and that success does require time. First was a Virtual Cockpit Development Program, where Microvision was the prime and Boeing the sub-contractor. They had won Phases I and II awards, an initial Phase III contract was signed in 1999, and additional awards came in 2000 and 2001 from the Army for flight-testing. One goal of the program was to replace all the gauges in helicopters with a helmet-mounted virtual cockpit, and the program had progressed to the stage of flight-testing.

Another success was the Advanced Adaptive Autopilot, an Air Force project under the Joint Direct Attack Munitions (JDAM) program. Guided Systems Technology was the prime contractor with Boeing the sub-contractor. Guided Systems worked with Phantom Works to develop this technology so that it could be incorporated into munitions.

A third success was the Cruise Missile Autonomous Routing System (CMARS) for the Tomahawk Mission Planning System. Scientific Systems Co.

was the prime contractor and Boeing the sub-contractor. Boeing became involved with CMARS from its role as the mission planning system prime on the Tomahawk project, and Boeing had worked with Scientific Systems since 1999 during Phase I and II contracts. This project illustrated the non-linear aspect of development, he said, with Phase II work starting in 1999 and a Phase III award from the Navy not beginning until 2004. Great patience is sometimes required to develop long-term partnerships that pay off.

He listed a series of questions that companies need to consider when working with the SBIR program:

- Is there a champion in the agency who can help from the beginning through insertion and implementation?
- Does the agency really want the technology, and will it accept it after development?
- Does the capability offer benefit at a system level?
- Does the benefit justify the transition costs?
- Can the prime contractor itself find champions for their programs, and also act as a champion for a technology being developed through Phase I and Phase II awards?

Advantages of a Team Approach

The funding issues discussed for agencies are relevant for the prime contractors as well, said Mr. Hendel. That is, companies need to find ways to fence off some money that can be earmarked definitely for Phase III projects. This was necessary to develop the technology not only through the Technology Readiness Levels (TRL) 4 and 5, for SBIR Phases I and II, but also to push the TRL higher to levels 7, 8, and 9 so it is ready for insertion into a prime contractor's program. In consulting with others at Boeing, he had heard suggestions in favor of a team approach in linking the small business, the prime contractor, and the customer early in managing the technology. Such interaction can prevent the isolation of activities in silos and promote collaboration.

He said that intellectual property (IP) issues, which concern many participants in SBIR programs, had so far not been an issue for Boeing as it worked with partner companies. The small companies owned the technology, and both companies worked on it.

He ended with two points of advice. First, he noted that small businesses with successes in SBIR Phase II did not approach Boeing on a regular basis to inquire about interacting with Boeing programs. Mr. Hendel said Boeing would welcome more such approaches. Second, he noted that a technology can have a negative impact on development logistics when it is inserted inappropriately downstream. He cautioned that this could sometimes be a reason for those working at a logistical support level to resist the insertion of a technology.

Mario Ramirez
Lockheed Martin

Mr. Ramirez, who is the officer responsible for small business participation on the Joint Strike Fighter (JSF) program of Lockheed Martin, opened his presentation by describing his company as a lead systems integrator and information technology company. Lockheed Martin does 80 percent of its business with the DoD and other U.S. federal agencies, and therefore, he said, "We certainly understand the urgency of establishing a corporate strategy to leverage in the SBIR program."

Currently, Lockheed Martin was in the process of establishing a task force to determine the current levels of SBIR involvement across its five business segments. The initial meeting was due to be held shortly, with the objective of taking the actions necessary to establish and integrate an overall SBIR strategy. He said that SBIR was an important component of the JSF program.

To make the SBIR program work, he said, data collection on new programs is critical. This process requires that the customer, the integrated product teams, and the supply chain collaborate to identify needs. These needs, in essence, determine the program's priorities and long-range needs. A second necessary element is the annual review cycle of technology, which leads to better opportunities to provide feedback in the overall SBIR process. Feedback is critical for both development cycles and integration, he said, and this approach enables the parties involved to align with long-range strategies and technology baselines for technology development.

Understanding the Customer's Needs

The company works with both capability roadmaps and technology roadmaps, drawn up by its engineers in partnership with the customer. This process gives Lockheed Martin a better understanding of the customer's needs, which is critical; enables the development of program priorities; and provides opportunities to integrate SBIR technologies into overall product roadmaps. Drawing up a complete corporate technology roadmap requires that SBIR is part of the picture. The company feels that SBIR awards of significant scope should be brought into product domain working groups and incorporated into the roadmaps, as appropriate. To make successful transitions to Phase III, SBIR technologies must be integrated into an overall roadmap.

Among the examples of what is working, he mentioned the Lockheed Maritime Systems and Sensors, which had 10 years of experience working with the Navy and partnering on SBIR technologies. The company planned to take such successes into account when conducting its analysis. Boeing had also had some early successes in the Joint Strike Fighter program, including the award of a Phase III contract for $6 million. They had also done well at integrating their supply chain and had captured an award at a sub-tier supplier for $100,000 plus options. He was proud of these early successes and predicted more.

A Need to Improve Procedures

At the same time, he saw procedures that could be improved. These included a need for strategic technology portfolios to assess strategic planning and provide clarity on reform acquisition needs. Also needed was better insight into the activities of laboratories, which at times competed with one another. He said that the goal for each laboratory should be to focus on its strengths. Once those are well known, it would be possible to provide a more systematic approach to communicate and share SBIR technologies throughout the company's engineering community—a critical step in assessing the company's needs in relation to the topics available.

Also, technology transition must be well coordinated and must include the customer, the supply chain, and small businesses. This coordination, he said, should also include advanced technology demonstrations, which should be used to integrate multiple SBIR awards into a complex weapons system. By brokering half a dozen such topics, advanced technology demonstrations could offer significant insight into the challenges of integrating these topics into a major weapons system. Too much leveraging of the advanced technology demonstrations, however, could make programs less risk-tolerant.

Sharing Responsibility

Another key element was how best to share responsibility. Lockheed Martin's government partners had many ongoing responsibilities, and SBIR was only one of many tasks. He said that the program could benefit by allowing a prime and/or supplier to share those responsibilities and offer the partnership as a technical point of contact.

Another issue that could be improved he called "produceability." That is, when an SBIR technology is judged to be ready for Phase III, a concern is not only whether the technology is sufficiently mature, but also whether the small business can produce it in the quantity required to sustain production. In addition, does the small business have the capital to make the significant investment required to support production?

Finally, although Lockheed Martin had not had difficulties with SBIR partners in assigning rights to intellectual property, the entrance into a Phase III contract would be the time to review any IP issues that need to be addressed.

He reviewed several procedures that might be adapted for use in the SBIR program. One was the Navy Advanced Technology Review Board's process to evaluate across programs to produce more effective transition of new technology. Also, he said he would like to create a version of the Joint Strike Fighter Science and Technology Advisory Board (JSTAB), a high-level S&T board that reviews programs' priorities. The JSTAB team consists of the program office, the contractor team, and S&T organizations of every service partner. Team members review technologies and establish priorities. He said that during the upcoming business segment analysis, when the company's five business segments will be evaluated, these two programs would be reviewed for lessons that might be applied to SBIR.

Developing a More Strategic Outlook

The fundamental challenge to improving the program, he said, was to develop a more strategic outlook. This would include a focus on long-term results, which is not always popular with the small business community. To maximize market impact, small business innovations must be aligned with the needs of government and the prime contractor; doing so can lead to more rapid and collaborative development of new technologies via technology mining. He said that Lockheed Martin had begun to regard this aligning process as an opportunity to engage more small business concerns. This process could be advanced by scouting small innovative research firms at the Navy Opportunity Forum and the DoD Phase II conference, for example, and by a greater commitment to outreach. This outreach should include the company's small business liaison officers, the technology leads, and business development specialists to produce an integrated effort that can address the concerns of small business.

Building a Relationship between Prime Contractors and Small Businesses

Lockheed Martin also intended to build more formal business relationships with its small businesses, which are critical to successful Phase III transitions. This process must begin with joint visits to customers when both sides can discuss product discriminators, areas for further investigation and collaboration within Lockheed's own Independent Research and Development (IR&D) and Cooperative Research and Development Agreement (CRADA) technology culture.[27] These relationships would also help integrate the SBIR technologies and

---

[27]The Department of Defense IR&D Program is designed to promote communications between the DoD and industry to increase the effectiveness of independent research and development activities and to ensure effective use of IR&D accomplishments to meet defense needs.

A Cooperative Research and Development Agreement (CRADA) is a written agreement between a private company and a government agency to work together on a project. Created as a result of the Stevenson-Wydler Technology Innovation Act of 1980, as amended by the Federal Technology Transfer Act of 1986, a CRADA allows the Federal government and non-Federal partners to optimize their resources, share technical expertise in a protected environment, share intellectual property emerging from the effort, and speed the commercialization of federally developed technology.

firms, and allow Lockheed to demonstrate its successes and build formal partnerships.

To bring a project to technological maturity, Mr. Ramirez said, it is critical to have adequate funding on hand. When a technology at a TRL of 4 or 5, for example, must be brought rapidly to a 7 or 8 for transition to the warfighter, the contractor needs to be able to deploy a financial incentive rapidly.

In summary, he said, Lockheed Martin believes that SBIR collaborations are attractive across the corporation. Initial explorations had created synergies across the five different business segments, and the SBIR task force was seeking to ensure that the necessary support elements are integrated into the strategic plan. This process was evolving, he concluded, with the objective of integrating senior management, mid-management, and operational personnel. This integration is essential because technology acquisitions decisions are made at the intersections of these levels.

John P. Waszczak
Raytheon Company

Mr. Waszczak introduced himself as director of advanced technology and SBIR-STTR at Raytheon Missile Systems (RMS), in Tucson, Arizona. He began by saying that a good deal of consensus had already been built during the conference. He said he would add to the discussion by recounting the process he had followed at Raytheon and some of the lessons that had been learned.

Raytheon's SBIR staff had spent a good deal of time with John Williams of the Navy SBIR program and Douglas Schaffer of the Missile Defense Agency (MDA), attempting to spread the SBIR program across the other services and agencies with which Raytheon worked. He cited a substantial opportunity for not only Raytheon but also for the small business partners who stood to benefit from the $2 billion spent annually on this program.

Raytheon, which is divided into seven business units, had 80,000 employees and revenues of $20.2 billion in 2004. In addition to Raytheon Missile Systems, in Tucson, the company consisted of Space and Airborne Systems, Raytheon Aircraft, Integrated Defense Systems, Raytheon Technical Services Company, Intelligence and Information Systems, and Network Centric Systems. One of these divisions, Integrated Defense Systems (IDS), had been working formally with SBIR for about two and a half years. Raytheon Missile Systems had been involved for about one year, and the company was in the process of integrating SBIR relationships across the corporation and corporate offices.

His division, Raytheon Missile Systems, was interested in the high-tech capabilities of potential SBIR partners. RMS produces a substantial portion lot of the missile systems procured by the U.S. government and allied nations. RMS products include air-to-air systems, surface Naval air defense, and standard land-to-air missiles, and its activities are classified under areas such as ergonomics, guided projectiles, directed energy weapons, kinetic kill vehicles, advanced programs (Mr. Waszczak's group), and land combat.

Using SBIR to Leverage Emerging Technologies

He said that Raytheon was focusing its activities on SBIR and STTR in order to leverage the technology emerging from the DoD portion of those programs (about half the total program value), as well as technology from the other SBIR agencies. SBIR is an integrated part of the company's strategic plan to enhance supplier diversity; half to two-thirds of a typical program in which RMS participates goes to subcontracts, and more than half of the companies supplying technology to Missile Systems are "small."[28] Raytheon aligns itself with both large and small businesses to ensure that the company is well represented in strategic technologies. Small businesses, he said, represent the "technology engine" of Raytheon and of the country, so that the company needs to develop better ways of integrating SBIR/STTR technologies in order to deliver the best value to its customers.

Raytheon sees SBIR as an extension of its R&D program. In Missile Systems, the ratio of development spending to research spending is about 3:1, and a goal is to find and maintain the best balance. While the Technical Director focuses on Internal Research and Development (IRAD), Mr. Waszczak focuses mostly on outside R&D, spending about 90 percent of his time on SBIR or STTR. He is looking for more opportunities for rapid technology development and insertion by establishing long-term relationships with key small businesses and strengthening relationships with customers by helping them get the right technologies to the warfighter quickly. He said that his goal was to better coordinate the activities of the government, the small businesses, and prime contractors like Raytheon.

The Need for Integrated Roadmaps

To move toward this goal, Raytheon Missile Systems worked with the Technical Director to make sure that an engineer's technology roadmap includes not just IRAD, but also other areas of R&D, including SBIR and STTR. "That's very important and critical," he said. "We're not funding engineers just to go to IRAD or a program office unless they have an integrated plan and can show how all the pieces fit together."

Each product line vice-president is named a "lead," including Mr. Waszczak who is the lead for Advanced Programs. The lead's goal is to interface effectively with the customer on technology roadmaps and to ensure that the company is

---

[28]The Small Business Administration defines a small business as a business employing fewer than 500 people.

working on the right technologies. He said that in expanding RMS, the product line leads provide direct links to the program managers in government. The "pull" from program managers or program executive officers is a key to technology development, as well as to effective research. The program offices are backed up by other functional groups, such as engineering. These functional groups drive the execution, based on guidance from the program offices.

This organizational system, developed within Raytheon Missile Systems and extended to Integrated Defense Systems, was now being expanded across the rest of the business units. At the corporate level, the vice-president of technology coordinated the leads that had been identified at each unit.

He then discussed "key entry points" to the SBIR process, from a prime contractor's point of view. Raytheon emphasized the entry point of Phase II to form relationships with small firms. But being involved in Phase I and Phase II, while offering near-term opportunities, was not always sufficient, he said. The company must not only discern what is being done now, but also what is about to be done. This stage of proactive involvement he called "Phase Zero," the time to identify the technologies and projects about to be funded, allowing the company to prepare for future opportunities as well as present ones. (See Figure 16.)

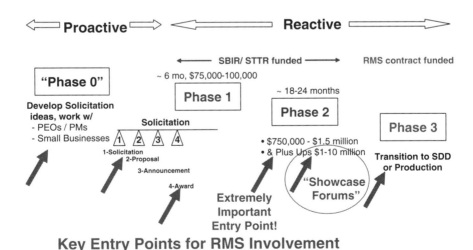

Key Entry Points for RMS Involvement

Raytheon evolving from Reactive to Proactive

FIGURE 16 SBIR key entry points.

Benefits of SBIR and STTR to Raytheon

He listed many benefits available to Raytheon Missile Systems by working with the SBIR and STTR programs. Among near-term business advantages were the abilities to help program managers solve problems, reduce costs and risks, and find alternative solutions. They worked with the advanced program managers and directors for each product line—the people concerned with tomorrow's needs rather than just today's deliveries and customer requirements.

Other benefits came from the SBIR's "phase transition" emphasis where RMS helped small businesses develop new components and worked with new programs to help integrate mission systems. In addition, RMS benefited by exposure to more acquisition candidates.

Box A Keys to a Prime's Success in Leveraging SBIR Technology

- Effective collaboration between government, small business, and prime.
- Work with government project managers to achieve "program manager pull."
- Make business case through focus on strategic technologies.
- Show how the SBIR adds value for company.
- Communicate effectively, internally and externally.
- Develop streamlined, user-friendly IT processes.
- Make SBIR part of company and customer tech roadmaps.
- Identify lead people in each product line.
- Identify lead people for key technology areas.
- Show how metrics flow downstream.

He showed some of the success metrics he had developed for this conference, with the key metric technology that advances to Phase III. He also focused on technologies that could be considered strategic and those on which RMS worked closely with a small business. One goal was to make sure that Raytheon's engineer on a particular program was responsible to the program manager within a small business, that the requirements were understood, and that the partners were working in coordinated fashion. For two important metrics—SBIR/STTR technologies leveraged and support contracts from SBIR awardees—RMS goals were exceeded by more than 100 percent. Raytheon had engaged three dozen small businesses in Phase I, two dozen in Phase II and, the key metric, three dozen in "Proposals open," the threshold to Phase III. Each transition to Phase III would mean creation of a new program and a return on Raytheon's investment.

Raytheon Success Stories

He said there had been many "success stories" of SBIR projects in which RMS had participated. A summary included the Mark 54 Torpedo Array Nose Assembly, with Materials System, Inc.; the EKV (Exo-Atmospheric Kill Vehicle), with Vanguard Composites; guided projectiles, with Versatron (now part of General Dynamics), the Navy, and the Army; and dozens of Phase I and Phase II projects across Raytheon. All of them were considered strategic technologies, not commodity-related projects, he said, because of the limited resources available. He also mentioned the success of about half a dozen "mentor/protégé" relationships ongoing within Raytheon. Successful mentor/protégé relationships with small high-tech firms represented potential SBIR-STTR working partners for the company.

He showed a diagram of SBIR processes and interfaces he described as "a complicated spaghetti chart," making the point that each time a prime like Raytheon works with a small business, the two partners should mutually create a roadmap to guide the development of the technology. Raytheon had developed a model that began with sharing capabilities and technology roadmaps for missile systems or concepts; then discussing with the customer the program evolution and enabling technologies required; identifying technology gaps; working with customers to develop program roadmaps; and moving the technology through the different SBIR phases, from Phase Zero to Phase III. Those SBIR awards can provide RMS with a valuable link to the overall program by providing the strategic technology enablers needed, as well as a competitive advantage or "discriminator."

Mr. Waszczak moved to a list of major issues facing prime contractors, and some recommendations to deal with them:

- **Lack of Efficient Links to Small Businesses.** Prime contractors need more forums to permit effective matchmaking across all organizations. Examples include the Navy Technology Assistance Program, PEO-IWS (Integrated Warfare Systems) matchmaking, PEO-W (Strike Weapons and Unmanned Aviation) solicitation requests, teaming with the Missile Defense Agency (MDA), and gap analysis with small businesses. Raytheon attended showcase forums like the Navy Forum and the DoD National Forum, where they had the opportunity to meet with small firms. These forums also provided benchmarks to measure how well Raytheon was doing in technology development compared to other organizations and agencies.

- **Inadequate SBIR Database for Awards and Solicitations.** Unless these data are updated more often than every 6, 9, or even 12 months, opportunities to interface with contractors receiving those awards are missed. The databases should be up-to-date, searchable, and organized by common standards.

- **Cultural Differences.** Prime contractors, government, and small businesses must all make culture changes to make more effective use of SBIR. *Prime contractors* must educate their leadership, which Raytheon is attempting to do, seek a diversity of suppliers, and overcome the "not invented here" syndrome. *Engineers* need to be alert for technologies being developed by others. *Government agencies* must involve prime contractors early, in "Phase Zero." Agencies must also plan early for the transitions from Phase II to Phase III, and encourage and educate small businesses to contract prime contractors for support. *Small businesses* need to recognize that the real payoff is in Phase III, not in Phase II. This realization should increase their motivation to form partnerships with prime contractors.
- **Hesitancy on the Part of Small Businesses to Work with Prime Contractors.** A leading cause of hesitancy is concerns over data rights. Mr. Waszczak said that Raytheon was working to educate small businesses about rights and to alleviate their concerns—for example, by using the proper MDA non-disclosure agreement clauses. This, he said, is essential for building trust between prime contractors and small businesses. Prime contractors also have to understand and communicate the unique provisions of SBIR, which protects the technology for the small business forever, whereas many MDA agreements last only three to five years.
- **Insufficient Cases Demonstrating Return on Investment.** Prime contractors cannot yet demonstrate enough successful transitions to Phase III. To do this, they need to integrate SBIR into the normal course of business and to continue their process improvement effort.

To complement this list of major issues, he summarized some of the recommendations that flowed from Raytheon's experience with the SBIR program:

- **Ensure Variety in the SBIR Matchmaking Process.** This can be done by sharing SBIR best practices across services, emulating the Navy's TAP Forum for matchmaking and taking other steps described above.
- **Emphasize Program Manager Pull.** This is key requirement for both existing and advanced programs, and should emphasize integration of the roadmap planning by government, prime contractors, and small businesses.
- **Focus on Program and Technology Roadmaps Gaps.** Again, firms must incorporate the SBIR as a key element of technology roadmaps. Timing is critical, because the primary opportunities are created by beginning early—in "Phase Zero."
- **Beware of Too-high Expectations of Small Businesses.** An early start may create unrealistic expectations, which can be costly and demoralizing. All parties need to emphasize the three-way matchmaking process.

- **Begin Early to Plan for the Transition to Phase III.** This requires setting aside adequate funding.

Mr. Waszczak summarized by restating Raytheon's commitment to the SBIR program as an "important part of our technology development process" and a key element of supplier diversity. The program helped fill "gaps" in the company's integrated roadmaps, which was the focus of current efforts. He was also working to fit SBIR into the company's business development toolkit, and working with engineers throughout the company to engage their participation. He closed by saying that he welcomed additional opportunities to communicate this message to others and to expand the SBIR program throughout his company

Earle Rudolph
ATK

Mr. Rudolph, the vice-president for strategy and programs at ATK Mission Research, said that he would describe the SBIR from a different perspective. He said he had worked as a government program manager, where he interacted with small businesses; at a major prime contractor; and at the Draper Laboratory. As a result, he said, "I understand the transition and the retention of intellectual property, and how that strikes fear into a small company that's trying to protect itself." Now, he said, he was in "a company that's in transition."

ATK had originated as a spin-off from Honeywell, best known initially as Thiokol, manufacturer of solid rocket motors for the Space Shuttle. ATK's core products also included conventional munitions, but the company was now in transition and had two sets of customers. The first was the DoD, and the second was the prime DoD contractors. "I have to understand where we're going, where they're going, and match my technology roadmap to both of them. Then we can use programs like SBIR to find niche markets to expand into." Planned growth markets for the company included hypersonic air-breathing systems, thermal resistant materials, advanced propulsion systems, time critical strike weapons, advanced projectiles, Advanced Anti-Radiation Guided Missile (AARGM) derivatives and directed energy. The company had made 11 acquisitions in the past five years, and hired more than 200 new PhDs.

SBIR Brings Access to High-level Expertise

Mr. Rudolph said that his company valued the SBIR program for many reasons. A growing $3 billion company, ATK is able to gain access to high-level technical expertise that would normally be beyond its reach. The company had used its acquisition process to expand or increase its technical sophistication, allowing it to create partnerships with prime contractors and with government that it would not otherwise have. The SBIR program augmented its growth strategy by enabling it to find and engage the kinds of entrepreneurs who have been "road warriors" or technical pioneers, often out of the industrial mainstream, who discover a technology and then look for a place where they can develop it further. This was often the "phantom works" or strategic group of an existing corporation. Having such people helps ATK nurture the new technologies needed by their customers at the DoD or the major prime contractors.

Strategies for Engaging Technologies

ATK has two strategies for engaging technologies. One is partnering. For Phase I, the company engages as a subcontractor in order to understand the technologies, assist in maturing the technology to a Phase II level, and establish trust and working relationships with the small firm. For Phase II, the company becomes a major sub (49 percent) during hardware implementation, testing, and maturation of the technology to a TRL of 5 or 6. For Phase III, ATK becomes the industry partner with the government as customer, employing discretionary resources for development.

The second engagement strategy is acquisition. ATK had made strategic acquisitions that were, for them, "fairly high-risk." The business strategy is to identify potential candidates with the SBIR process, conduct technical and business due diligence, and make a decision to buy, with the key to the decision being the retention of intellectual property. Successful examples included the AARGM, which came out of an SBIR II contract and led to purchase of the company three years ago— at considerable financial risk for a company the size of ATK. The acquisition of IP rights involved much discussion and has since resulted in case law to support the process. ATK also purchased Mission Research Corp. (MRC) in a different type of acquisition, emphasizing multiple leading-edge technologies and bringing to ATK a fast, agile and entrepreneurial technical base. Successful SBIR programs at both companies convinced each side to do the acquisition.

Advantages of SBIR . . .

From the company's perspective, the SBIR program has several advantages. It allows second- and third-tier firms like ATK to have a competitive advantage. These firms can leverage their discretionary funding with technologies already identified as relevant by the customer. Firms the size of ATK do not have a large engineering base, so the SBIR forums hosted by the services offer a valued alternative to engage with engineering expertise. He cited the recent Navy forum as providing a place where small firms, prime contractors, and government customers could engage in dialogue, exchange and even active negotiations. In addition, the program provides incentives for entrepreneurs to take risks and form new alliances in order to develop innovative solutions, essentially broadening the national R&D base.

. . . and Impediments to Execution

The SBIR program also has impediments, he said. First, the administrative processes are too slow to allow small business owners to move their technology forward quickly. This frustrates the small business owner and, from the point of view of the prime contractor or subsystem-level contractor, unduly slows the movement to market. From ATK's perspective, one reason that transitions fail is that the SBIR program often focuses on technical solutions to the neglect of systematic planning. He cited AARGM as a "systems answer to a systems problem," which was why it has done well. But if the firm and the customer do not decide what the system is and how the technology fits that system, even an excellent technology may not prevent a business failure.

He closed by reiterating the real concern of the small business for its intellectual property and the retention of IP rights. The protection of rights and positions is essential, he said, both for their confidence and for the relationship with a prime. If a prime is to work with a small firm as partner and help take that partner into production, the smaller partner must retain some level of control over the intellectual property. Only when the larger partner pays attention to the needs of the smaller firm will both participants realize the potential advantages of the partnership.

DISCUSSANT

Trevor O. Jones
BIOMEC

Trevor Jones thanked the speakers for their presentations, and said he would like to raise several issues he had not yet heard discussed. First, he asked whether any agencies had thought of taking an equity interest and/or options on small businesses, or given small business partners convertible equity loans to fill the funding gap. Second, he raised the possibility of encouraging industries to create their own equivalent of the SBIR program—as commercial ventures within companies, but ventures with a short timeframe. At present, he said, the gestation period from Phase I to Phase III is often too long to maintain the interest of a commercial entity. He asked for feedback on both these possibilities.

Mr. Waszczak said that Raytheon had several times experimented with total acquisition of small businesses. He was not sure whether it had done any partial equity transactions, but that they would certainly consider such an approach. Mr. Hendel estimated that Boeing had done three total acquisitions, but did not know of any partial equity arrangements.

Making Good Use of Roadmaps

Trevor Jones then offered several impressions. First, the program managers in the agencies sometimes served a valuable function as "match-makers" between prime contractors and small firms. At other times, the prime contractors with technology needs might be the matchmakers, finding a capable small business to take to the program managers.

Mr. Waszczak said that he has shared some very good exchanges about technology with program offices and Program Executive Office groups. They had then taken the further step of comparing technology roadmaps with that of their primary small businesses to make sure that all three of the organizations were in synch.

Mr. Hendel said that Boeing also used technology roadmaps to track the firm's projects and needs and set priorities for the coming year. He said he knew from his experience with the Multi-Mission Maritime Aircraft Program that programs in his organization worked closely to identify technology needs and topics that could become SBIR projects.

Mr. Ramirez said that Lockheed and JSF integrated small business concerns into their overall roadmapping process, opening significant opportunities to SBIR awards.

Trevor Jones asked if a subcontractor to an SBIR award winner who has an idea and a product can ultimately become the contractor who sells that product to the agency. In some cases, their work will result in a production product—especially when that organization is the sole source for the procurement. Mr. Williams responded that the Navy had done so with one SBIR company. After the small firm completed Phase II work, the Navy formed a partnership with the firm, which then developed and sold the product on a sole-source contract. He said the Navy's database had been reviewed and searched for companies to which the Navy had awarded contracts or some type of work. It was not easy to identify all firms in the database that had done SBIR-related work, but the number of such firms appeared to be large.

Trevor Jones also asked about first supporting development of a technology to the stage of commercialization, then issuing a request for proposals to learn whether other competitive technologies existed. Mr. Waszczak said that Raytheon, after working with a small business, usually would know that their technology was the best, most cost-effective solution. At the same time, the company would indeed continue to test the marketplace to be sure that this was true.

DISCUSSION

Mr. Kidalov then invited discussion from the panel members.

Incentives for Contracting with SBIR Firms

Mr. Kidalov said he had heard that even inside a large company SBIR firms needed a champion, a corporate strategy, and incentives for the company to continue using SBIR firms, even beyond the competitive advantages they provide. He asked whether or not the panelists saw value in a system that would allow for recognition of efforts to contract with SBIR firms, perhaps from Congress and the agencies.

Mr. Hendel said that when the agencies award contracts to prime contractors, incentives are built into the contracts. He said it should be possible to offer the prime contractors similar incentives for working with SBIR contractors or development projects.

Mr. Ramirez said that incentives are critical to technology transitions, and would stimulate additional competition and more SBIR-type technologies and companies.

Mr. Waszczak said that for Raytheon an important incentive would be to streamline and otherwise optimize the SBIR process, which would ensure the development of many technologies needed for the long term. A second incentive would be assurance that customers have realistic plans to support the transition from Phase II to Phase III. Third, companies all have requirements to work with small and disadvantaged businesses, and SBIR relationships would help meet those goals.

Mr. Rudolph pointed out that individual business units, like agency program managers, need to see value in what they do, and dislike the risk of new technologies or small companies without a track record. They need incentives and other encouragement to take these risks.

Mr. Waszczak repeated from his presentation that metrics are important in any aspect of the business, including measurements across the industry and across the SBIR process. Such metrics might be specific goals for industry, or a more general goal to take SBIR technologies into Phase III.

John Williams of the Navy SBIR program reminded participants of the importance of having and implementing incentive and risk-reduction strategies. The DoD, he said, has been promoting spiral development, technology insertion, and similar steps, but he found that funding for technology insertion work was often deleted from acquisition programs when overall program funding is constrained, since both prime contractors and DoD acquisition managers are risk adverse—and new technologies are inherently risky. He asked two sets of questions:

- Should the review committee recommend that DoD acquisition programs set aside money to perform technology insertion work?
- How can the DoD measure prime contractors' Phase III subcontracting activity with SBIR firms, how should the DoD create incentives for this activity, how is technology assessment ("due diligence") paid for, and

how does the DoD ensure that SBIR insertion work is properly budgeted and what steps can be taken to protect those funds?

How to Finance Phase III

Mr. Waszczak addressed the question of how the government should provide funding to help SBIR firms make the transition from Phase II to III. He advised against tapping the fund of set-aside money, preferring to have program managers realize the value of SBIR activities. Steps to encourage acceptance include basic education about how small businesses can be technology engines for important technologies. Managers also need to realize that it can take three or four years to bring new value, unless a project is on a fast track, and that not all SBIR firms will make it into Phase III. Once managers do see the added value of the program, he suggested, the transition process would begin to take care of itself.

The Need to Educate Program Managers about SBIR

Mr. Rudolph agreed on the importance of articulating the value of SBIR to the program office and the Program Executive Office. A small company cannot be expected to do this, because it is focused on developing and explaining the technological aspect of its work. He had found it useful to sit down with the technical staff and show them how a technology would be used, which helps the staff to develop the right technology. This, in turn, leads to the buy-in of the program managers and the program executive officers.

Mr. Hendel agreed that education for program managers was needed, so that they see the need for the SBIR program and understand how it can improve performance and lower cost. Only then will they develop pools of money that could be utilized on a regular basis for Phase III awards from a prime. He called it "an education process by the government to us, and by us internally, and when we get to a certain point, the processes all fall into place and happen naturally."

A Changing Role for Prime Contractors?

Dick Reyes, president of a small technology company, raised the question of changing the environment that had made it possible for the large prime contractors to dominate his market space. He recalled that in 2003, the top 100 DoD firms had 89.9 percent of the total federal R&D budget, with Boeing and Lockheed Martin together accounting for more than half. He asked the group whether they thought it would be desirable and possible to change the contractor environment, or whether this would be blocked by the large prime contractors in their desire to dominate the marketplace. Or, is it the responsibility of the DoD to bring about change? Why would the prime contractors change unless the government forced them to change?

Mr. Waszczak said the SBIR program was part of a larger cultural change affecting all three entities—the government, the prime contractors, and small businesses. He said that for most companies, the future would bring more horizontal versus vertical integration. The prime contractors were getting out of the business of building and designing everything. Two-thirds of their costs were now going out to suppliers, and because half of those suppliers were small businesses, the prime contractors were motivated to take advantage of that technology engine and work with them. "We don't see you as competition 95 percent of the time," he said. "We see you as enabling technology to allow us to bring the total system to the government."

Mr. Rudolph added that five years ago, Lockheed, Boeing, Raytheon, ATK, and others would not have gathered as they had today to discuss how better to deal with SBIR and small business. Such a dialogue would not have occurred, or it would have occurred only at a governmental level. In addition, he said, having been a supplier to the larger prime contractors, "I can tell you they're interested in diversifying their supplier base so that one single supplier does not become the single point of failure. I work mightily to get around that attitude, like you do."

Accelerating Innovation:
The Luna Innovation Model

Kent Murphy
Luna Innovations

Dr. Murphy began by saying his firm was in the business of "driving innovations to equity, creating actual corporate value." His firm had built several successful businesses, has a continuous pipeline of opportunities, and pursued the objective of "accelerating the whole innovation process."

He said that he began his career as a teenager, working as a janitor at an ITT laboratory. While at the laboratory, he was introduced to the nascent field of fiber optics, and by age 19 had learned enough to earn several patents for fiber-optic telecommunications components. Recognizing this talent ITT allowed him to work not only as an inventor, but also to build manufacturing equipment that helped develop these inventions into marketable products.

He then earned a degree in engineering at the Virginia Polytechnic Institute and, while completing a master's program, invented a fiber-optic sensor that was licensed from Virginia Tech even before the university had a technology transfer process. He published a paper, and several large defense contractors began buying the devices. Later, Dr. Murphy and his partners began performing contract R&D work for the defense contractors and then for other Fortune 500 companies. At that point they learned about the SBIR program and won their initial SBIR awards, which helped move their products more rapidly into the marketplace.

Reflecting on their business model of moving innovation to the marketplace, Dr. Murphy and his colleagues realized that they did not have to invent everything they needed themselves. They realized that the laboratory shelves of universities and federal laboratories across the county held an undeveloped backlog of interesting technology. They knew that there was a wide gap, however, between those inventions and the ability of university professors and scientists to write business plans and find corporate sponsors or venture capitalists to fund these ideas to commercial fruition. Recognizing this opportunity, they built a network of researchers and developed their company into what is now Luna Technologies. Their work was supported by licenses and patents from products developed through R&D contracts and equity in the spin-off company themselves.

Today, Luna today consists of:

- **Luna Technologies.** This parent company is headquartered in Blacksburg, Virginia, where it was formed, and has divisions in Roanoke, Danville, Charlottesville, and Hampton Roads, and, most recently, northern Virginia. While the locations in south and central Virginia attracted little venture capital, even during the "bubble days," the SBIR programs helped the company grow and create hundreds of jobs in rural parts of Virginia where high-technology employment is scarce. The parent company employs more than 135 full-time scientists and engineers who work on contract R&D projects. It also funds 40 to 45 full-time people at Virginia Tech, most of them experts in materials and integrated systems. Dr. Murphy noted that he spent most of his time in what he called the "Luna Triangle," bounded by Baltimore, Maryland; Blacksburg, Virginia; and Research Triangle Park, North Carolina. This area, he added, supports some $15 billion in federally funded research. About a third is done at the top research universities and two-thirds at federal laboratories. He said that his company had much to offer in moving the worthy technologies produced by that research into commercial products.

- **Luna Innovations.** This first spin-off company grew out of a technology developed at NASA Langley. It was brought into Luna Technologies, which combined its own money with an SBIR award to build a prototype technology that was then sold to Lucent. Even with this customer acceptance, Luna was unable to find funding for the model it was building, so it commercialized a simple product for the telecommunications market that the financial community could understand. It raised two rounds of venture capital, totaling $12 million, and built a line of telecommunications products that competed with products from Agilent, JDS Uniphase, and other large companies.

- **Luna Energy.** This spin-off is based on a technology that was part of Dr. Murphy's original master's thesis. Dr. Murphy and his group steadily developed the technology with early funding from Boeing, Northrop Grumman, and Lockheed Martin. Luna later won SBIR awards from the Air Force to develop strain gauges, and from NASA to develop skin friction balances based on this new technology. Even though NASA products typically have a small market, the skin friction gauges caught the interest of the oil and gas industry because of their potential for use in harsh environments—the high pressures and temperatures of deep wells. These energy companies invested $12 million to build a product line, and eventually bought the organization, named Luna Energy, and located a division in Blacksburg, Virginia.

- **Luna *i*Monitoring.** Founded in 2002, Luna *i*Monitoring, developed beta prototypes of harsh environment wireless sensors that won SBIR Phase I and Phase II awards from the Navy. While the firm was working to integrate this technology into the Navy, it was acquired by HIS Energy, an information-handling company that is part of a $20 billion European conglomerate. This Luna spin-off, given $2 million in operating capital, $1 million in up-front capital, and a 10 percent royalty stream on the product line, moved into an abandoned warehouse in Roanoke, where it manufactures these products and continues to expand the product line.

- **Luna Analytics.** The founding technology for this spin-off was discovered "on the shelf," this time at Lucent, which had developed it for the telecommunications industry. Luna was interested in using it for life sciences and won SBIR awards to develop and build prototype systems. The products were sold to life sciences companies to study protein-protein interactions, which led to an agreement with a biotechnology firm to keep the company in Blacksburg as it continued to introduce products commercially.

While Luna does some basic research, its main emphasis is to tap into federal and commercial markets and use market-driven knowledge to educate the university and federal laboratory partner about potential markets and their needs. It also works with many corporations that have intellectual property that they would like to develop for the market. Luna takes those ideas and tries to fund them through R&D contracts with either federal or commercial partners. After the proof-of-concept stage it seeks additional funding, which is usually supplied by venture capital, corporate partners, or internal investments. (See Figure 17.)

Luna's objective is to accelerate the innovation process in this way; the outcome may be a spin-off, a stand-alone company, a licensing agreement with a bigger partner that has better access to markets, or a product that Luna keeps at the parent company. As the company continues to grow, it accumulates more expertise internally, both at identifying markets and the tech transfer process, including how to raise financing and how to build sales and marketing channels.

Dr. Murphy illustrated how his products move through the "technology flow pipeline" from basic research to applied research to prototype to product, by describing several current programs:

- **Flame Retardant Additives.** Luna responded to a Phase I opportunity for flame retardant additives by gathering unique data during Phase I that led to a Phase II award and then a prototype. This prototype consisted of a simple composite panel of carbon fiber and resin to which a fireproofing polymer was added. Unlike most flame retardants, which emit toxic smoke

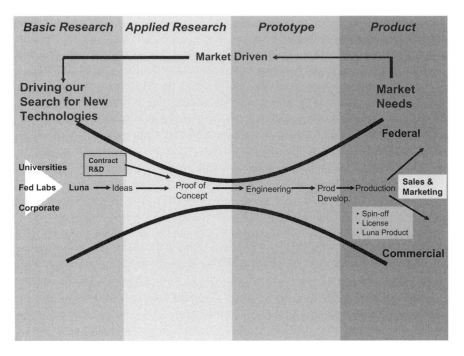

FIGURE 17 The Luna business model.

when burned, this additive emits only a low-toxicity smoke and is suitable for use with bedding, bed clothing, and other fabrics.

- **Luna nanoWorks.** Luna has been working with nano-materials for four years, using a trademarked base molecule called a "trimetasphere." This is a carbon-80 "cage" with three metal ions and a nitrogen atom at the center. The key to the technology is to place Gadolinium chelates securely inside this cage, which mitigates the reliability and safety problems that have beset current nano-technologies. The company hopes to use this technique to develop a more effective, safe, and durable contrast agent for use in magnetic resonance imaging. Another hope is to achieve the long-sought goal of safe cell targeting to combat cancers and other diseases. Currently, cell targeting has not succeeded because attaching the cell-targeting molecule to the outside of existing contrast agents allows the Gadolinium to remain inside the body long enough to be toxic. Luna has demonstrated the ability to securely attach binding molecules to the trimetaspheres.

Dr. Murphy emphasized the value of building not only several successful businesses but, more importantly, maintaining "a great continuous pipeline of opportunities and a way to utilize the research already in the federal labs and the universities and to help them move it into the marketplace more rapidly." He said that an important tool was the flexibility to use multiple mechanisms for funding, including SBIRs, venture capital, corporate partners, and profits from Luna's work.

In summary, he emphasized the need to pay attention to the whole innovation process, not just segments of it. Within his organization, some people enjoy the research world more than the products world, he said, but in every meeting, Topic A is always "where are we on the research, development, and product pipeline. There's always a need for basic research, but there's also a need to make sure we're pushing all these things toward a market." This, he concluded, is what is meant by accelerating the innovation process.

DISCUSSION

Dr. McGrath began the discussion by noting that the role of venture capital had barely come up in the context of SBIR. He noted that several of Luna's spin-offs had attracted venture funding, and asked whether venture backers had owned more that 50 percent of the company, which would make them ineligible for SBIR.

Dr. Murphy said that of Luna's spin-offs, only one did not qualify for SBIR, but none had actually sought an SBIR award. Once the firms found venture capital or a corporate partner, they had focused on the technology around which they had been formed. He said he had mixed feelings about using venture capital. Small VC firms should not be disqualified from the SBIR process, he said, but he was not sure that the largest investment firms, such as GE Capital, had a role. In such cases, he saw potential difficulties with ownership, control, and organization size.

Dr. Gansler said that distinguishing between operating companies that have a venture investment group and those that were "pure" ventures was now under debate, and would be one of the issues addressed in the Academies' studies.

A questioner asked what the panel thought of (1) offering a 20 percent tax credit for angel investing, and (2) changing the fiduciary rules regarding foundations so they could take higher risks on angel/seed capital investments. Dr. Murphy agreed that both suggestions had merit, and said that he supported other creative suggestions for strengthening American innovation, such as tax incentives for purchasing capital equipment. He said that many industries had outdated capital equipment, but felt strong pressure to value their quarterly profits more highly than investing in the company's long-term future.

SBIR as a Factor in Luna's Success

Dr. Wessner asked whether the Luna model was likely to have developed without SBIR support, or ATP support.

Dr. Murphy replied, "Absolutely not. There's no way we would have built the company that we've done today, and had the successes that we've had, without both the ATP and SBIRs." He said that he had tried "multiple times" to raise venture capital, but that the only two venture capital investments they received came just at the end of the late-1990s "bubble" days. Since then, he said, the dollars invested by venture capitalists had declined sharply, and venture capital companies were interested only in firms geographically located near their own offices. One venture capital company had expressed interest in investing in Luna's nano-technology company—provided Luna would move the operation to Silicon Valley. Luna had also had offers to sell operations to companies located outside the United States. Only the SBIRs had allowed them to keep the operations and jobs where they were.

The Need for Patience in Developing Technologies

James Rudd of the National Science Foundation congratulated Dr. Murphy on his company, which had some NSF funding for magnetic resonance imaging (MRI) work. He asked about the amount of time small business people should expect to spend developing an invention and introducing it successfully to the marketplace.

Dr. Murphy said that the time required varied by technology and by market. For example, their development of skin friction gauges took 10 years, from the time of discovery when he was a master's student to the time it was proven useful in oil wells. From the time the technology was proven useful, it took another two and a half years to complete the sale to oil and gas companies. In the case of the MRI contrast agents, four years of work were necessary to produce amounts sufficient for animal and other testing, and another four years would be necessary to receive FDA approval, produce the product, and generate revenues. Dr. Gansler added that some software products might be developed in 18 months, while a new vaccine might require 18 years.

The Innovation Continuum

Kevin Wheeler of the Senate Committee on Small Business and Entrepreneurship staff asked for suggestions to increase the number of Phase III awards. Dr. Murphy suggested better use of market-driven research—evaluating market needs, especially in the Defense Department, and communicating those needs rapidly to the researchers. Ms. Wheeler asked if Luna placed greater emphasis on its innovation or on its business activities, and Dr. Murphy said that Luna saw the two as parts of a continuum, and the company's mission was to facilitate the entire process. He described a need to sharpen the whole technology innovation process, bringing market information back to the researchers as rapidly as possible and helping to guide them. Even though not all discoveries and inventions end up as products, Luna tries to select the most promising technologies more quickly and reduce the cycle time. Ms. Wheeler also asked whether there was a difference between STTRs and SBIRs in the Phase III stage. Dr. Murphy said that awards differed by agency, and even by individuals within agencies, and changed over time, so a general answer was difficult to give.

Dr. Wessner added that Dr. Murphy's firm is a remarkable example of the interaction of regional strength and federal support. Because it is locally rooted in areas where market funds are not likely to reach, its success owes a great deal to the SBIR program. He also raised the issue of the 20 percent tax credit for angel funding. While this incentive might benefit many small businesses, such as restaurants, few angel investors were attracted to the complex high-tech areas needed to strengthen and revitalize the industrial base and offer high-tech employment. The SBIR and STTR awards are more specifically designed for high technology firms.

Challenges of Phase III:
SBIR Award Winners

Moderator:
Kevin Wheeler
Senate Committee on Small Business and Entrepreneurship

Ms. Wheeler briefly introduced Panel III, saying it would feature successful companies with experience in surmounting the challenges of Phase III.

Anthony C. Mulligan
Advanced Ceramics Research, Inc.

Mr. Mulligan said that his company had been founded in 1989 with capitalization of a thousand dollars and hope of a Phase II award from the Navy. His group had to survive for six months until a $500,000 contract came through in 1990, and the firm had grown steadily since then.

Advanced Ceramics Research (ACR) has two manufacturing facilities as well as a sales office facility in Washington, D.C., and has plans for two more manufacturing plants. Based on contracts in place, total projected revenues for 2005 were about $23.2 million, he said.

The business has evolved away from its early dependence on SBIR sales. For 2005, about $2 million of sales were projected to go to commercial, non-government customers; about $17.2 million for non-SBIR customers, primarily military; and about $4 million in the SBIR program.

Of the company's total sales over its 16-year history, about a quarter will have been recorded in 2005. About $22 million came in commercial sales, primarily to the computer hard-drive industry, and $36 million came as non-SBIR government sales. About a third of the government sales were R&D transition dollars to take SBIR programs to commercialization. (See Figure 18.)

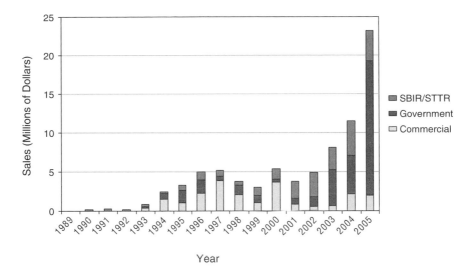

FIGURE 18 Sales history.

In total, the company had won almost $23 million in SBIR awards, averaging about one-and-a-half or two Phase II awards per year. They had won 75 Phase I awards and 25 Phase II awards altogether, most of them within the last two years. "We are on a very fast growth track," Mr. Mulligan said.

He noted that commercial sales had dipped in 2001-2002 because the company moved most of its commercial manufacturing to a plant on the Tohono-O'odham Indian reservation, near Tucson, Arizona. Those sales are now growing rapidly.

Mr. Mulligan described the company's commercialization strategy, emphasizing the following strategies:

- **Perform Work that Is Core to the Company's Strategic Plan and Have a Clear Path to Commercialization.** Unless this path is clear, his company will not write a proposal. The company had, on rare occasions, won Phase I awards that they turned down because their strategic plan had changed. They had also won Phase II awards or been asked to submit Phase II requests and turned those down because they did not meet the core strategic plan.
- **Hire the Best Possible Talent.** The company is looking for cum laude graduates, PhDs, and people who are very competitive. They have found that employees with a built-in will to win are easier to manage, which fits well with the company's aversion to micromanagement.

- **Apply Effective Incentives.** The company's bonus plan is four pages long, outlining how to get bonuses and how to earn more in many ways. For example, bonuses may follow a patent application, a patent award, or the sale of a license or commercialization of a product. Employees also share in license royalties and fees—everything important to the management of the company.
- **Emphasize Commercialization at Every Opportunity.** The management of the company is always focused on this message.
- **Reach Out to the Warfighter.** Previously, the company spent most of its outreach time with the program managers who control the budget in their effort to win government sales. Now they try to determine what the warfighters want and need, then "push the middle" to help that happen.
- **Constant Training.** The company takes every opportunity to advance employees' education, including sending people to school and hiring consulting teams.
- **Reach Out to Congress.** The company tries to educate its Congressional members about procedures that are and are not working well. Although a difficult task for a small business, this has been important in helping to narrow the gap to Phase III transitions.
- **Leave No Stone Unturned.** Perhaps the most important element, he said, is to take advantage of every factor in trying to leverage their assets and maximize their opportunities.

As an example of the last point, Mr. Mulligan described the commercialization of a technology at ACR that was completed in the early 1990s. It involved a set of ceramic composite technologies that originated from a Navy SBIR program for computer-automated control. The topic tied in also with a NASA program and a DARPA program, allowing ACR to develop a new technique of precision machining that revolutionized the manufacture of computer hard-drive disks. The technique brought down the price of the disks so dramatically that one out of every two hard drives in the world has been manufactured with the ACR technology. From 1991 to 1999 the company sold about $14 million worth of this technology.

SOME PHASE III TRANSITIONS

He also described two recent case histories of Phase III transitions, both of which came quickly. The first was a water-soluble tooling technology the company began working on in 2002. This was a 6.1 or basic research program originated by the Office of Naval Research to develop the science as a way of making cheaper parts for the Joint Strike Fighter (JSF). During Phase I, this program won a prestigious R&D 100 award and ended up with a $25 million ID/IQ[29] contract

---

[29]Indefinite delivery/indefinite quantity.

from NAV-AIR in 2004, of which $5 million had been funded and $6 million was due by contract. The technology had proven versatile—it was able, for example, to build inexpensively a part like a wheel, with hollow spokes and hubs, which could be made previously only with expensive tooling. This elicited a significant agreement with the automaker BMW, which wanted to use the technology to produce sport-utility vehicle parts, and has roused interest from both VW and Audi for the same purpose.

The second case history concerned the small unmanned air vehicle (UAV) business, which began in 2001. The Office of Naval Research granted a Phase I STTR in 2001, which had transitioned to $17.6 million in total sales, only $6.5 million of which were SBIR-STTR contracts. By the end of 2005, total contracts and sales are estimated to be nearly $30 million. He showed a chart of UAV business progress, including different UAVs and different technologies in UAVs, especially sensors, which would probably provide nearly half the company's revenues in 2006.

He then listed a series of lessons learned:

- **Hard Work and the Desire to Commercialize Do Not Guarantee a Successful Phase III.** There are still real barriers in the acquisition system.
- **The Navy's Technology Assistance Program Can Help Small Businesses Integrate a Complex Process.** ACR had done four SBIR programs with them, of which the first three had already succeeded in commercializing, and the fourth seemed to be moving toward success. He suggested a careful look at what makes the TAP so effective.
- **While Breaking into the Federal Acquisition System Is Difficult, ID/IQ Contracts Ease and Accelerate the Process.**[30] ACR had three ID/IQ contracts, two at the Office of Naval Research and one at NAV-AIR. Both places, he said, made it easy for those who want to make purchases quickly.
- **Program Managers Need Incentives to Work with Small Businesses.** Program Managers in the federal acquisition community do not intentionally shun the small business community, but they have no strong incentive to embrace a new technology or process from a small business when the risk is likely to be higher.

In closing, he said that the SBIR program works very well. It is highly competitive, and ACR wins only three out of every ten programs it tries to win. However, said Mr. Mulligan, the acquisition system is falling short in not being ready

---

[30]An ID/IQ contract is a contract between a federal government agency and a contractor for the indefinite delivery of an indefinite quantity of services. Timing and delivery of the services is determined through agency completion of an individual task order or individual delivery orders.

to take advantage of fast-moving technologies and capabilities. There is no effective bridge between the acquisition community and those who are developing innovative technologies. Building this bridge, he said, should not be the sole responsibility of the SBIR community; the acquisition community must help pull those new technologies across.

Nick Karangelen
Trident Systems

Mr. Karangelen introduced himself as president of Trident Systems, a small systems engineering firm. He was a Naval Academy graduate in the class of 1976—"one of [Admiral Hyman] Rickover's boys in the nuclear Navy during the Cold War," serving in submarines. He then went to work at TRW for a couple of years, and then for IBM to work in Manassas, Virginia, on submarine combat systems.[31] In 1985, he said, "my mother died, and I decided that life is short and I better decide what I really want to do." With about $11,000 in his pocket he started Trident Systems, which has grown today into a $25 million company.

Trident had produced a number of successes including bringing in about $30 million in revenue over the last 12 years from touch-screen technology developed under SBIR contracts as well as achieving a number of DoD transitions for SBIR technology. He said he would not talk about these successes, however, because he wanted to focus on how to improve Phase III of the SBIR program.

Improving the Acquisition System

He referred to the end of Mr. Mulligan's talk to open a discussion of the shortcomings of the acquisition system. Mr. Karangelen agreed with the need to address these shortcomings and began with the FY 1999 Defense Authorization Act. That Act outlined six procedures the SBIR program should follow. The first was that major acquisition programs should designate liaisons to the SBIR community. This had been done, he said, but had only designated individuals in laboratories that represented dozens of programs. What was needed, he said, was a designated SBIR liaison for *every* major program. Another Defense Authorization Act recommendation was to establish good linkages between SBIR solicitation topics and acquisitions people. This, he said, was also done, and today some 60 percent of SBIR programs are directly related to acquisition programs.

---

[31]The Command, Control, Communications and Intelligence (C^3I) contract for the Virginia-class submarine was awarded to Lockheed Martin Federal Systems of Manassas in April 1996. This system is noted for its extensive use of open system architecture and "COTS" (commercial off-the-shelf) components, many of them produced by small businesses, bringing a new degree of system affordability and flexibility.

A third recommendation was for senior acquisition executives to issue guidance to acquisition programs that would make SBIR part of their ongoing program planning. The goal was for these executives to keep track, as they moved to the next milestone, of which SBIRs were in their plan and, most importantly, in their budget. With the exception of a few "aggressive, innovative, and enlightened" program managers in the Navy, and perhaps a few elsewhere, this was not being done.

He said that an explicit directive to include SBIRs in the mainstream of acquisition was written into the revision of SBIR documents[32] by Dr. Gansler, but it was removed before finalization. He said that other Congressional reports over the years had also contained "good ideas" that were never implemented, most likely because they would mean changes to the way "business had always been done."

He blamed not the large DoD prime contractors, but the DoD itself, which had grown comfortable in dealing primarily with big prime contractors and large, horizontally integrated companies to address acquisition needs. This focus on the largest companies had led to less competition, even though many medium-sized companies, such as ATK, which offered good alternative technologies, were disappearing. It was also hard for the smaller companies, such as Trident and ACR, with just a few hundred employees, to survive in the current environment. He said that the DoD, was resisting change, as "all big, bureaucratic organizations do," and that the push for change needed to come not only from Congress but from the top of the agency itself. He added that even Dr. Gansler, who was Defense Under Secretary for Acquisition and Technology in the Clinton Administration, was unable to change the system from within.

Moving SBIR into the Mainstream

Mr. Karangelen urged the conference participants to "take a hard look at what it will take to change the ratios"—that is, to reduce the 89.9 percent of all federal R&D contracted out was presently going to the 100 largest firms. He said that less than four percent of the R&D budget contracted out was contracted to businesses of fewer than 500 people.[33] He also said that of all R&D spending of every kind generated by the federal government, only 0.4 percent goes to small technology businesses, even though perhaps a third of all scientists and engineers in the United States work for small businesses.

---

[32]This directive was placed at section 5000.2R of the 5000-series documents of the SBIR program.

[33]The SBIR program generally defines a "small business" as one employing fewer than 500 people. This adheres to the guideline of the Small Business Association for manufacturing and mining firms. Other "small businesses," concerned with trade, construction, retail, agricultural, and other activities are usually defined in terms of average annual revenues.

He said that the fault did not lie in the SBIR program itself, which was highly competitive, drawing dozens of firms to compete in every topic area for Phase I awards. The problem was that the acquisitions system did not consider the SBIR to be in its main stream of activities, so that program managers seldom reach out to the small firms, even though they represent a huge resource of talent and innovative energy. Nor do program managers always see the advantage of the SBIR program in helping them to vet out small companies able to work with government contracts and identify the best among them.

A Suggestion for Filling the Funding Gap

Mr. Karangelen suggested that better acquisition habits could go a long way to fill the perceived "funding gap." That is, by the time a program manager finishes a Phase II competition, several small companies might be vying for the same technology area. Normally, one of these could be chosen with confidence for a Phase III contract. If the firm is not ready, a good practice would be to issue a smaller version of a Phase III contract that would allow the company to survive as it develops the technology. He argued that the funding gap "exists largely in our minds, and in the minds of the program managers," because the money is there, but it is usually committed to a large prime contractor already under contract. In all, he estimated, some $400 billion in potential funding was tied up in this way and unavailable for funding the gaps between Phases I and II and the acquisition system.

Unleashing the Innovative Power of Small Business

Mr. Karangelen said that his basic message was that there was nothing wrong with the SBIR program. It worked well, especially in the Navy program where it had been "put on a pedestal and polished." The fault lay instead with those program managers who did not consider it to be part of their mainstream. If these program managers would place the SBIR in their budgets and in their planning, he suggested, they could unleash "the innovative might of small business in America" and the country would reap a bounty of "better products faster and cheaper."

He closed by quoting the 1982 law authorizing the SBIR program. Its purpose was "to ensure that federal R&D procuring officers and program managers make use of the wealth of resources available from small businesses in addressing the mission and research needs of their agencies." The SBIR program was not a welfare program for small businesses, he concluded, but a way to get the DoD what it needs "faster, better, and cheaper."

Thomas Crabb
Orbitec

Thomas Crabb, vice president and chief financial officer of Orbitec, said that his firm was founded in 1988, around the same time as other companies represented on the panel. He and his early partners began, he said, as "three guys in their houses, working in the back bedroom, trying to make things happen." The initial focus was to develop technologies for plant research in closed systems for the International Space Shuttle.

Within a few months, the founders had won several contracts and the firm, based in Madison, Wisconsin, evolved rapidly. Orbitec now employed nearly 100 employees, and revenue had grown from $6.3 million in FY03 to an estimated $11.8 million for FY05. Their main customer has been NASA, along with the DoD and a few other agencies. The firm was approaching $100 million of Phase III awards. Although getting these awards was difficult and time consuming, an even greater challenge turned out to be maintaining NASA funding once contracts had been signed.

Among the technologies and products produced for government were high-quality flight systems, high-performance combustion and propulsion, portable simulation for training and operations, and innovations for systems and components. Among commercial products being pursued were environmental controls and tissue culture for the biotechnology markets, LED lighting for aquariums and human lighting applications, instrumented biomedical devices, plasma processing, nanocoatings, water purification, and environmental sensing and control. The firm also planned to do "next-generation agriculture" to allow crop growing in manufacturing warehouses.

Looking Beyond SBIR to Commercialization

The strategy for dealing with the federal government within Orbitec was to arrange the firm's capacities—engineering analysis, electrical systems, and mechanical software—toward current contracts as well as beyond them. In this strategy, government Phase III awards are regarded as "transfer points" leading to other markets beyond government procurement. The firm also maintained a pipeline of technologies that are ready to approach commercialization, all of which share the ability to integrate into single systems for the Space Station or Shuttle and therefore have very low power needs, and small mass and volume. Such constraints, he said, require and breed new technologies.

In order to deal with both technology push and market pull, Orbitec had created a separate entity called Planet LLC, which is a technology incubator. Product development and licenses flow from Orbitec to Planet, while royalties, R&D for product upgrades, new market-pull ideas, SBIR marketing plans, licensing agreements, and commitments for SBIR flow from Planet to Orbitec.

Orbitec's involvement with SBIR Phase I and Phase II awards began during the firm's first few years. Beginning work on its first Phase III around 1994, it encountered frequent budget problems starting with the first award in 1998 and the second in 2000. However, because it did good work in the Phase III, it received more Phase III work. "So Phase I and Phase II awards played a very, very good seed capital role in establishing Phase III," he said, "which has really been the engine of growth." The company had experienced many different combinations of grants and outcomes, with Phase I grants, for example, leading to additional Phase I awards, Phase II awards, and even directly to Phase III commercialization.

The Advantage of "Small Phase III" Grants

He discussed Orbitec's "key NASA Phase III experience," in which NASA had been helpful in providing a "small Phase III" to get over the Valley of Death that loomed beyond Phase II. The technology was not ready for a full Phase III, but the agency gave the company time to demonstrate it with a smaller grant, enabling it to fly a demonstration payload. With that success, it was rewarded with a Phase III contract of $57 million for the "development and flight of a Plant Research Unit for the International Space Station," followed by additional contracts. Without that initial "push" to help the firm survive the funding gap, he said, many subsequent products would not have been developed.

Mr. Crabb added that this success was enabled by the unbundling of a large NASA contract in 1995. The contract was for a complex life sciences system called the Centrifuge Accommodation Module, which included an incubator, cell culture unit, insect habitat, and other habitats. Two large aerospace firms, Lockheed and McDonnell-Douglas, were competing for the whole contract, and Orbitec was a subcontractor to McDonnell-Douglas, which was working on the Plant Research Unit. NASA decided to divide the contract, rather than awarding it all to a single, large, prime contractor. This permitted some contracts to European firms and allowed three assets to be procured by small businesses, including Orbitec. The small firms gain invaluable experience, as well as funding, by contracting some of these large jobs, which would not have been possible without the contract unbundling.

The Downside of Contract Downsizing

He brought up the subject of contract downsizing, in which a proposal is accepted by a customer but proposed budget levels are not met. He said this was a serious issue for the SBIR program, with a long history. Orbitec's $57 million contract for the Plant Research Unit—its largest NASA contract to date—was reduced by more than 80 percent after the first year. After the second year it was again reduced, by more than 78 percent. The company survived only because of the lobbying efforts of allies in Congress and at NASA. Orbitec endured similar reductions on a contract for an animal habitat. In all, he estimated that the company would lose more than $60 million on Phase III contracts. These reductions were not due to Orbitec's performance, which had been rated excellent by their customer within NASA.

He attributed this to his company's lack of sufficient clout within NASA to maintain budget levels at a time when NASA was changing its priorities, notably including a shift in emphasis away from the International Space Station and toward the moon and Mars.

The impact of Phase III contract reductions was especially severe because the contracts were by then the growth engine of the company. He showed a slide illustrating the impact of the reductions over a five-year period, when the Phase III revenues grew from a trivial level to the majority of revenues.

The Good, the Bad, the Ugly

He concluded by summarizing Orbitec's experience with Phase III. The "good," he said, was that the Phase III process screens and selects worthy projects. Phase III advances good businesses that meet government needs, enables small businesses to become prime development agents for agency needs, and can fuel commercialization through the Valley of Death.

The "bad," he said, was that Phase III can last more than five years before a technology is ready for commercialization; Phase III opportunities are not frequent, especially where large, bundled contracts preclude opportunities for small firms; and few key people are aware of the benefits of Phase III to government.

The "ugly," he added, was that a small firm cannot count on contracts from Phase III once won, bringing skepticism from investors. Small business activities seem easier to cut from budgets because they have less clout within the federal government. From the small firm's point of view, one result was a "poor credit rating": "The bankers and the investment community need to trust that that contract is going to be there. You can't have a contract cut by 80 percent and have your financiers trust that you can maintain your technology development."

He recommended several steps for improving what he considered "an already great program":

- **Continue Efforts to Unbundle Contracts.** This will enable small businesses to have more opportunities to participate.
- **Solidify Government Commitment to Phase III Projects Once Awarded.** This should include a Phase III termination clause with enhanced payments and penalties, and some guarantee of proposed funding within 20 percent of contracted amounts.
- **Apply Incentives to Motivate Large Prime Contractors to Partner with Small Businesses.** This can including penalties for avoidance or re-

engineering and rewards for strategic efforts to build a culture of joint working relationships.

- **Provide Bridge Funding to Incubate Phase III Projects.** Once transitioned, he said, the benefits of small business should be clear.

He closed by recommending several strategies for small businesses interested in participating in the SBIR program. First, create a product strategy that transcends the government need and fits commercial markets. Second, never write a Phase I proposal without having a vision for the Phase III and/or commercial product. And third, maintain advocacy activities throughout the SBIR work to seek allies among technical monitors, ultimate users, contract managers, SBIR management, and congressional representatives.

Robert M. Pap
Accurate Automation Corporation

Mr. Pap said that he would review his experiences, "both good and bad," with SBIR projects, which extended back to the founding of his company in 1985. Accurate Automation, with 22 employees and projected 2005 sales of $3.5 million, had won 71 Phase I awards (4 STTRs) and 46 Phase II awards (4 STTRs) and had produced 17 Phase III products and contracts. The firm designs, develops, and manufactures emerging commercial technologies, such as unmanned boat and aircraft systems and signal processing devices.

To illustrate some of the company's projects, he began with a video representation of the cockpit of a jet fighter. At the request of the National Transportation Safety Board, the company had developed the ability to compress an image and validate the image without using a watermark or altering the image. He then showed the bridge section of an aircraft carrier. Using its first SBIR in 1985, the company developed the ability to send data from several places on land to an aircraft carrier. That ability had evolved into a new product designed for use with radar. Finally, he showed photos of three unmanned vehicles, part of a family of unmanned boats, airplanes, and other systems for the Navy.

SBIR Projects

He then discussed the X43A Telemetry Subsystems, a $1.5 million Phase III project for the Air Force, which had evolved out of the X30 National Aerospace Plane project. During the X30 program the company was asked to bid on a project that the larger contractors preferred not to do—fault diagnosis on the fuel distribution system. The fuel in this case was slush hydrogen, which is highly explosive and dangerous. Accurate Automation was able to design a safe telemetry system to transmit data to the ground in real time, even during the X-43A's destruction in June 2001. It was then used successfully on later non-crash flights as well.

Next he described the Phase III MALD SDD, a unique 150-to-300-pound-thrust jet engine required by the Navy as an SBIR. The MALD was a new design for an air-launched guided/powered flight munitions that could be dropped as a decoy from a B-52 or an F-16 aircraft through an F-35 bomb bay. Accurate Automation was encouraged by the Navy to prepare a bid and won a small contract from the Air Force, assembled a team, and bid on the larger MALD program. The team designed a full-sized system and built a prototype capable of advanced autonomous flight that could mimic a real rocket as well as do obstacle and threat avoidance.

A Project with an Unhappy Outcome

The experience did not turn out well, said Mr. Pap. Their engine, although developed specifically for the requirement of a decoy, was not considered, even though it was able to thwart the ability of a sophisticated enemy to detect its decoy nature through jet-engine modulation. The engine was not being used in the system adopted by the Navy.

The more important part, he said, was the payload. His company developed a technology of plasma dynamics for the Army, Navy, Missile Defense Agency, and NASA. It protected the payload area from the body of the aircraft, and allowed it to carry payloads that were dangerous to the rest of the vehicle. This meant that it could operate not just with a radio on the front, but could carry a directed energy weapon, which was important to the United States.

"Unfortunately for us," said Mr. Pap, "we were teamed up against Boeing, Northrop Grumman, and Raytheon. We spent a million-and-a-half dollars, put together a team that competed with everybody in the industry, and the winner, Raytheon, will not even talk with us."

The Air Force selected Raytheon, bypassing Accurate even though it was the low bidder. Certain individuals, he said, were at that time giving contracts to various prime contractors, and "we were victim." But the company built the vehicle, and sold one to NAV-Air, and the technology was used on an unmanned boat that the Navy was using on mission modules for a combat ship.

As a result of this experience, Mr. Pap offered the committee several recommendations:

1. When an SBIR company and a major manufacturer offer a bid, and the bid is led by the SBIR company, the past performance of all the major companies should no longer be allowed as a selection consideration.

2. When the selection authority has used false statements in the Source Selection Decision Document to award a contract to a major company over

the SBIR company, the SBIR program office should be authorized to run a parallel effort as a Phase III that is supported from the DoD element line item of the program. Each modification to the contract would be provided to the SBIR-led team until it has been given a satisfactory Phase III award by the selected prime.

3. Lastly, SBIR program offices in DoD should have funds to support (under contract) a major proposal by an SBIR-led team when existing hardware has been developed by SBIR funds. Unlike the large prime contractors, a team of small businesses has no funds to compete in a major program through the "plus-plus reimbursement in their overhead."

Protecting Sensitive Knowledge

He raised the issue of protecting sensitive knowledge. He had been approached by the Navy, a major supporter, which in the mid-1990s wanted to fund the company's work on plasma. Accurate Automation developed a device that protects a wave guide from an HPM or EMP attack.[34] Accurate Automation now has a contract and has tested the device extensively for duty on ships. He noted that the company tried to work with a university on the project, but could not reconcile the need for weapon secrecy with the university tradition of publishing research results.

He described a similar problem with another technology, a radio frequency mitigation device to protect radar and electronic warfare systems from HPM and EMP attack. Descended from work begun in 1979, this technology acts as a shield to divert an incoming shell. Again, he said, such a "disruptive" technology, which fits the SBIR model, must be developed in secrecy. It is not well suited for university research, both because of the secrecy requirement and because it takes 7 to 10 years to develop.

He closed by urging a continuation of the SBIR tradition of supporting high-risk research, and addressing questions of interest to the "best and brightest" PhDs of all ages. He also urged more of the "skunk-works" approach, where technologies for the military could be developed under conditions of intense focus and, where necessary, security.

---

[34]Electromagnetic Pulse (EMP) is an instantaneous, intense energy field that can disrupt at a distance numerous electrical systems and high-technology microcircuits that are especially sensitive to power surges. A large-scale EMP effect can be produced by a single nuclear explosion detonated high in the atmosphere. This method is referred to as High-Altitude EMP (HEMP). A similar, smaller scale EMP effect can be created using non-nuclear devices with powerful batteries or reactive chemicals. This method is called High Power Microwave (HPM). Source: Clay Wilson, "High Altitude Electromagnetic Pulse (HEMP) and High Power Microwave (HPM) Devices: Threat Assessments." Congressional Research Service, April 2006. Accessed at <*http://www.stormingmedia.us/47/4787/A478744.html*>.

Mark Redding
Impact Technologies, LLC

Mr. Redding said his company was founded in 1999 to work in an area of business known as "predictive equipment health management technologies," which included conditioned-based maintenance (CBM) and prognostics and health management (PHM). Since the company's formation, it had been awarded more than 45 SBIR Phase I contracts and 27 Phase II contracts, with customers including most DoD agencies. The company received the Tibbetts National Award in 2002 for demonstrated SBIR success, and its first Navy ID/IQ contract, in 2004, for $25 million.

The company was formed with a staff of five people in 1999, and by the end of 2004 it employed 56. By the middle of 2005 it had reached 75 employees, mostly mechanical, electrical, and software engineers. Revenue growth had roughly paralleled employment growth, with projected 2005 revenues of $9.3 million. The number of SBIR awards in the past year had been relatively flat, while revenue and employment continued to grow, which one would expect during a transition to Phase III. That is, some technologies were moving into commercial markets and supplying additional revenue. He noted, "We would not be the company that we are today without this program."

A High Success Rate

Impact Technologies had earned contracts from many major agencies, including Navy, Air Force, Army, DARPA, ONR, AFOSR, and NASA. Its commercial defense customers included Honeywell, GE, Boeing, Goodrich, Rolls-Royce, Lockheed Martin, Northrop Grumman, and Pratt & Whitney. Compared to the national average success rate on Phase I proposals of about 10 percent, he said, the rate at Impact had been over 50 percent. In converting from Phase I to Phase II, where the national average success rate is about 40 percent, the rate at Impact was approximately 90 percent. He attributed this success at least partly to the practice of teaming with either a large prime or a university in the Phase I proposals and in clearly identifying the customer's needs at the outset. The primary focus within the company was now transitioning or commercializing the SBIR-developed technology.

One of the technology areas of Impact was equipment for detecting and diagnosing equipment problems and predicting the future operation of that equipment, which allows maintenance to be scheduled at opportune times. The company does this for a broad range of applications, including avionics, propulsion, AMAD/drive train, structures, sensors/data, flight controls, and fuel/hydraulics.

Other systems where the company develops technology are the Joint Strike Fighter (F-35), CH-47D Chinook Helicopter, H-60 helicopter, USS Briscoe with

Gas Turbine CMB, DD(X) submarine, Expeditionary Fighting Vehicle, M1A1 Abrams Tank, FCS Manned Ground Vehicles, and F-15/F-16/F-117 aircraft.

The company had recently been awarded a small Army contract to analyze wartime data from an Apache helicopter to use as justification for diagnostic systems on such equipment. Impact was able to react quickly to that need, and to actually perform risky development of technologies while not under contract. He said that a large company would be unlikely to do this without a contract.

Improving the Phase III Transitions

He then offered some observations, based on the company's experiences, regarding the Phase III transition challenge:

- Impact had found Phase III funding to be very limited. While the company had won more than 30 contracts considered to cover Phase III, the dollar amounts of the awards were relatively small, averaging about $50,000. He believed that this was because the company was not part of the acquisition process.
- The technology readiness level at the end of Phase II is generally not sufficient to allow commercialization. This creates a funding gap between completion of prototype development (Phase II) and actual insertion or commercialization. The $25 million ID/IQ contract mentioned above was funded with an initial delivery order of $50,000—the only funding that had been awarded to date.
- There was no clear funding path or insertion policy for SBIR-developed technologies.
- The company's experience in selling to the large prime contractors had been difficult. Unclear issues included a "not invented here" attitude and unclear allocation of intellectual property. The most nettlesome issue for the company had been the difficulty of meeting the standard contract terms and conditions of a large prime. He told of being selected after competitive bidding and then having to wait six months for the small firm's data rights clause to be inserted by the prime into the contract.
- The company's most positive experience had been the Navy's Transition Assistance Program. "It's by far what I would consider to be the best example of helping small businesses commercialize," he said. "It's not the final answer—I think improvements could be made—but it's a good first step."

The Need for Overall Improvements

He proposed several improvements for SBIR. The first, which would be inexpensive, would be to educate the large prime contractors about the SBIR program, including its objectives and some of the IP issues. He also proposed a new type of SBIR program, similar to the STTR program, with a new type of contract that requires a small business to be teamed with a large prime for the technology insertion. This would be, he said, a "real Phase III contract." Finally, he suggested additional funding for TAP-like programs, with more focus on networking and brokering deals between the small businesses and the large prime contractors. He noted that part of the Navy program includes a small amount of market research to identify potential customers, and urged more such research.

He reiterated that for SBIR to be a success, incentives were needed for the large prime contractors to integrate SBIR-developed technology—but only for technologies that are deemed to be worthy by the DoD customer. He suggested that financial incentives were needed to induce the large prime contractors to integrate those technologies.

In closing, he raised the issue of whether venture capital firms should be allowed to participate in Phase I and Phase II SBIRs. He opposed this because of the likelihood that large VC firms would soon use their own SBIR companies to compete against the other "true" small businesses. He repeated that much of his company's success had come as a result of teaming with large prime contractors in Phase I and Phase II, which had equal access to all small companies. Many of their Phase I proposals were supported by a large prime, such as Boeing, which also provided support to other small businesses. If such a firm had its own VC-backed company, it would be unlikely to support a proposal from Impact. And its own VC backed company, with its vast resources for proposal-writing, would be able to produce a better proposal than could a traditional small company like Impact Technologies.

Tom Cassin
Materials Sciences Corporation

Mr. Cassin began by noting that Materials Sciences Corp. (MSC) was a very small business, so that his duties reached all the way from negotiating bank lines of credit to filling in as forklift driver when someone failed to show up for work. His company had 30 full-time employees, with approximately $6 million in revenue. Of the 30 employees, 27 were engineers, leaving three people to answer phones, write contracts, and process checks. The company relied heavily on outsourcing for auditing, CPA, legal, and other needs.

MSC was formed in the early 1970s to pursue fundamental research in characterization of composite materials. In the early 1990s, under new management, the company refocused on engineering services and intellectual property development through teaming arrangements with other companies, spinning off a company or licensing a technology. Currently, the company was expanding and adding capabilities, especially in manufacturing and testing of material systems and specialty materials development.

The company is located in an industrial park, like many small businesses, and has a testing support facility on-site with limited production. Most of its work is done for the Army and Navy, along with some commercial contracts. It works on Naval structures, including the DD(X) IDHA, the CHSV, and the EFV/AAAV.[35] It has Army contracts for an Advanced Composite Bridge, a Modular Composite Bridge, and an Advanced Composite Military Vehicle. It also has contracts (such as COPV Life Extension and Structural Health) to test and analyze products to extend their lifetimes by using new techniques and ways to manage them.

Difficulties of Technology Insertion

He suggested that a common theme his company shared with other SBIR firms is technology insertion, or perhaps the "velocity of technology." He referred to the perceived 18-month life cycle of software and computer hardware, and compared it to the lifetime of a project involving basic materials development or materials integration, which would be 10, 15, or even 20 years. When a company develops a new materials system, he said, there is a barrier to entry, which is caused by an absence of design data, an absence of suppliers, and the real or perceived risk of any new product or technique, including the likelihood that it will be too costly to market. The good news, he said, is that once a materials system is accepted, it is likely to be used for many years, creating a strong incentive to push for technology insertion.

A Strengthening Alliance with Prime Contractors

He said that the SBIR program had been fundamental to his company's involvement in both adding new technology to the DoD and strengthening its alliance with prime contractors. Recently, the prime contractors had begun to approach his company with suggestions, such as, "We have a technology gap. We need a certain technology, but we cannot insert that technology while we're working on a program." Materials Sciences had the flexibility to explore a new technology, develop it to an acceptable level, and insert it into an ongoing program. In more than half a dozen cases, an SBIR-originated relationship with the government and a prime had moved beyond the SBIR process so that his company had become a partner in designing programs.

---

[35]DD(X) IDHA refers to the composite Integrated Deckhouse Assembly (IDHA) for the Navy's next generation destroyer. CHSV refers to the Navy's Composite High Speed Vessel. EFV/AAAV refers to the Marine Corp's Expeditionary Force Vehicle/Advanced Amphibious Assault Vehicle.

Challenges in Working with the Program

He then listed a series of challenges his firm had encountered in working with the SBIR program.

- **Overcoming Fear of Risk.** The largest challenge by far, he said, was convincing a project manager or Program Executive Officer to take the risk of incorporating a new technology from a small business. For the prime contractors, he said, risk drives everything, and he could easily put himself in the place of a project manager having to answer to upper management that a risky critical element in the design of a multi-billion-dollar program had been awarded to a small firm that might not be up to the task. He said that the challenge for both sides is to understand what the risks are and to reach a common understanding of how best to deal with them.
- **Protection of Intellectual Property.** Legal costs are not covered by an SBIR contract, which means that a company has to pay for them out of profit or whatever monies exist to patent the technology. In addition, the cost of enforcing a patent if a competitor is infringing on it can be a million dollars or more. And SBIR firms, whose results are public, are vulnerable to infringement claims on existing patents, and it is expensive to demonstrate innocence.
- **Managing Cash Flow.** The company must capitalize facilities and fund inventory and receivables. If a small firm spends half a million or a million dollars on a deliverable, and the technology is delayed by the recipient's shipping department for a month or two, and by receiving for another four months, there is no mechanism to recoup the cost of that delay.
- **Lead Time Required for Qualification.** Like many participants, Mr. Cassin criticized the long wait for SBIR decisions.
- **Second-Source Development.** A customer is likely to say that the firm has an excellent product, but wants to know where else they can buy it if the firm disappears. The firm has to facilitate the effort to reduce the customer's risk.
- **Creating a Sustainable Infrastructure.** In the specialty materials business, a firm may rely on one or two subcontractors or suppliers for hard-to-find materials. A prime may see this as another element of risk, so it must be mitigated in some way.
- **Product Diversity.** Mr. Cassin returned to the image of the Ferrari and the hay wagon. A company that parks its Ferrari to jump on a hay wagon—a partnership with a prime—may still not be safe. "The wheels may fall off; the hay may catch fire; Congress may decide a program is no longer needed. If something happens, the small firm has to run back to the Ferrari again and speed off in search of another hay wagon."

Financing Difficulties

Aside from the SBIR program, the small firm has few choices for financing. Both venture capital firms and government dilutes the company's control over its intellectual property. Even so, he said, his firm and other small firms were motivated by the opportunity to explore the unknown and create new, marketable technologies. He said there is an entrepreneurial spirit in America congenial to the small business and the adventure of a new technology. He invited the prime contractors to join in that spirit, which had begun to happen. "The primes are engaged," he said, "and it's just a breath of fresh air. There's early buy-in right now." The danger is that a small firm develops a good product and then finds no allies to help develop it. He was encouraged by the awareness on the part of the SBIR agencies of the Valley of Death, and said that getting through that valley depended on small businesses, not just the government.

He closed with several suggestions. First, in many cases there is just one known customer for initial SBIR-developed technologies. He said that the Navy TAP program helped his firm market its technologies to more customers. Second, he praised the models of OnPoint and In-Q-Tel as alternatives to government grants. The message he took from these programs was, "Invest in me, and you can get a return on your investment."

DISCUSSANT

James Turner
House Committee on Science

Mr. Turner began by voicing his enthusiasm for the work of the panels, and optimism about the improvement of the program. He noted that when he worked on the original SBIR act, the framers realized that there were two kinds of technologies that could be called "commercial." One is a technology designed for government use; the other is a technology designed for use in the private sector. He said that this difference was not stated explicitly during the discussions so far. For example, the teaming of small business with a prime is aimed at insertion, while teaming with a venture capital firm leads to private sector sales. He stressed the importance of remembering this distinction.

Adapting the SBIR Program to Changes in Business

He also recalled that in putting together the SBIR program and the amendments, "what we were doing was trying to think SBIR as a system—how you get from idea to prototype to actual commercialization." He confessed that the model used was somewhat "clumsy" and linear, pretending to move through prescribed "phases." Even so, he said, in the 20 years since the program started, many program managers have been creative in using and adapting the program, to the benefit of high-tech small businesses. The businesses described by the panel members were "completely different animals" than the first companies in the SBIR program, and "the framework has to change to accommodate that."

"Managing" Serendipity Well

He said that the SBIR program has to plan for serendipity as well—for the company that begins by selling pet supplies and finds itself with a high-tech product, or another that starts in aerospace and ends up in biological systems. "One cannot predict where a great idea will lead," he said, giving another example, of an entrepreneur who began to work on defense projects for display systems, but whose first commercial application was a technique to thwart counterfeiting of currency. For Phase III, it is important to think about the government as a whole, because the ultimate user of a technology is not easy to anticipate and may not even be in the agency that provided the original funding.

In the area of SBIR contracts, he warned against changes that have the appeal of reducing red tape, but that would also lower the protections for small businesses that are presently written into the Act. One such change would be to replace the term "contract" with the term "other transaction authority,"[36] which he saw as a worrisome possibility. "I don't think 'other transaction authority' is anything other than a contract that gets around a lot of the protections, such as intellectual property rights, which are in the existing act."

He then invited panelists to elaborate on ways to improve the commercialization process.

DISCUSSION

A Cure for Contract Termination

Mr. Crabb said that his company could probably secure funding, such as investment bank loans, based on Phase III contracts. This funding would help them get through the "valley" to the next stage. He suggested several steps. The first is to examine the termination clauses. A Phase III contract can leverage additional investment if the firm is able to put limits on the ability of the customer to terminate the contract. This would essentially stabilize or guarantee the con-

---

[36]For example, the Defense Advanced Research Projects Agency (DARPA) has had the authority since 1989 to enter into contractual arrangements called "Other Transactions" with its private sector R&D partners. Other Transaction agreements are characterized by enhanced flexibility and reduced administrative burden when compared with the typical government procurement contract. Congress granted ARPA this "Agreements Authority" in recognition that a procurement contract is not the appropriate type of agreement for every form of Government-supported science and technology project. *<http://www.darpa.mil/body/d1793/intro.html#FN(2)>*.

tract, allowing the company to reassure other potential investors. With the termination clauses now in effect, he said, there are no guarantees; it is a one-way contract, with the decision power held by the customer.

In addition, he urged the agencies to unbundle their contracts. He suggested that one way to do this would be to motivate R&D teams of large and small businesses to work together. Also, he noted that the practice of NASA to grant "little Phase IIIs" allowed his company to prepare technology for insertion, and urged this practice for the other agencies.

The Issue of Earmarking

Mr. Karangelen raised a new issue—additional funding received under the name of SBIR through congressionally mandated budget insertions. Many small firms receive funding by directly asking members for assistance. The firm may argue that the technology is sound, and that an agency customer wants it, but the agency has committed all its funds for the year. The member, after checking with the agency, may earmark several million dollars into the budget to fund the project. As a result, a number of the more successful SBIR Phase III funds do not come from the DoD program managers' initiative but from Congressional action. After some discussion of the pros and cons of this practice, he suggested that it would be preferable for the agencies, rather than Congress, to pick the best SBIR II outcomes and put them into the budget process with Phase III funding.

More about Risk

Dr. Parmentola added that agency S&T shared the same problem with transition as small business. That is, many agency research programs lack the technological maturity to please a program manager in terms of risk. The program manager has funding to test that technology but is reluctant do so where there is risk. In order to mitigate this risk, a program manager will look for people who have the experience of taking a concept all the way to engineering design, building a prototype, and entering commercialization.

A questioner asked whether Phase III activities are likely to attract the interest of the private capital markets, whether angel capital, venture capital, or other sources of investment funding. Mr. Redding answered that private capital is not interested in this area, especially in the aftermath of the "bubble." Government as a customer is regarded as too risky, because of its ability to withdraw from a contract if policies change.

The Argument for Open Systems

Dr. McGrath asked about open systems and open-system architectures. The Navy was actively developing these, he said, and trying to determine the best acquisition strategy, such as unbundling. Mr. Karangelen responded that small business was a natural ally of what might be considered an open system, because such a system could easily be partitioned, or unbundled. The large prime contractors have created monolithic systems that are not open, he said. An open system would be one in which a small firm could build a part of the system and integrate it without the prime's involvement. He said he was a strong advocate of open systems, being in the combat system business, "but it's a real struggle."

James Rudd of the National Science Foundation said that NSF was involved in commercialization, but mainly with the private sector rather than with the DoD. He noticed that large corporations used systems to effectively involve small businesses and whether those techniques might be applied by DoD. He cited Proctor & Gamble as a large firm that hired certain companies, such as Nine Sigma and InnoCentive, to help them identify the best SBIR company to provide a particular service. He asked whether such companies might be helpful in the DoD space.

Mr. Pap said that he received several SBIR awards from NSF and had worked with search companies. The difference for DoD is that the agency wants to be able to use the technology it funds for its own purposes, and often there was no equivalent user elsewhere. Dr. Gansler added that a goal of the SBIR program is to sell a technology to DoD or NSF, and also to sell it in the commercial world. DoD then benefits from the lower costs and more rapid innovation stimulated by the marketplace. There are barriers, however, that prevent the most effective combination of civil and military technologies, and he urged the participants to address these.

Best Practice for Agency Programs: Program Executive Offices and Program Offices

Moderator:
Peter Levine
Senate Committee on Armed Services

Mr. Levine said he has observed the SBIR program from the perspective of the Senate Armed Services Committee for the last 10 years or so, and three characteristics stood out for him. First, the SBIR program is a highly competitive program under which the DoD and other federal agencies fund private sector entities to perform science and technology research on behalf of the federal government. Second, at least within DoD, the program has been successful in developing technologies in areas of military need. Third, the SBIR program, despite its competitive nature and high rate of success in developing technologies, has been less productive in bringing those technologies to commercialization. There is a highly competitive front end, a highly successful R&D process, and a much less productive process of transition. He reported a high volume of complaints about the weakness of Phase III, largely from participants in the program.

Accordingly, he asked the panelists to focus on those characteristics of the SBIR program. He said that those three salient characteristics were not unique to the SBIR program, but could also describe, to a significant extent, the S&T program of the DoD as a whole, with its highly competitive program for identifying good ideas that could benefit the national defense. This program, too, was very successful in bringing ideas forward and developing them, but success dropped off when it came time to field them.

He asked the panelists to consider whether the barriers to fielding technologies from the SBIR program were different from the barriers in the DoD as a whole. He said that the distinction was important because it would affect what remedies might succeed. Is it simply a question of doing a better job of technology transition across the Department, he asked, or are there unique problems with the SBIR program that need to be addressed?

Richard McNamara
U.S. Navy

Mr. McNamara, Program Executive Officer for PEO Submarines, described himself as an advocate of small business, and said that the centerpiece of his advocacy was the SBIR program. In his Requests for Proposals (RFPs) he incentivizes primes to subcontract certain percentages of the work to small business. For example, he contracted with General Dynamics on the Virginia-Class Program demonstrating that small businesses are a high priority and offered a million-dollar "bounty" per hull as an additional incentive fee for contractors who met small-business sub-contracting goals. The Navy owes it to the large prime contractors, he said, to provide real incentives for a policy considered truly important.

Advantages of an Outreach Strategy

He said that attending meetings such as this one was part of his outreach strategy to share SBIR experiences and promote the program. He said that he spoke at conferences for many groups and hosted conferences for women-owned businesses. He also visits laboratories and other activities to suggest how they can get more actively involved in SBIR. In actual SBIR transactions he said he has dealt with about 150 different companies over the past decade. Of those, he found that about one in ten was a company he would take anywhere, on any job; one in ten he would not recommend; and the rest were reasonably competent firms that had not reached the transition stage.

He said that a significant feature of SBIR companies is that "they're new faces on the landscape. People don't know them." He said that many people do not have the confidence to put money into a Phase III with an SBIR company, but his experience has given him the confidence to take that risk. In doing so, he found that the benefits outweighed the risks and that SBIR awards have become his preferred way of bringing competent small businesses and new faces into the submarine contracting community.

Suggestion in the Gansler Memo

He recalled a memo of SBIR suggestions written in 1999 by Under Secretary Gansler and said that his office has followed many of them.[37] For example:

---

[37]August 10, 1999, Memorandum from Under Secretary Jacques Gansler on the SBIR Program. The memo requested the assistant secretaries of the Army, Navy, and Air Force, the Acquisition Executive of the U.S. Special Operations Command, and the Directors of the Ballistic Missile Defense Organization, the Defense Advanced Research Projects Agency, and the Defense Threat Reduction Agency to, inter alia, "issue guidance to your Component's acquisition program managers to include SBIR as part of ongoing program planning, and to give favorable consideration, in the acquisition planning process, for funding of successful SBIR technologies."

- **A Program of Advocacy.** PEO-SUB advertises the SBIR program within the Program Executive Office and Team Submarine, which employs about 600 people, through a program of active advocacy. He educates program managers that SBIR is a tool for them to use, a way of recovering the tax dollars set aside for the program in ways that solve problems for them. Program managers compete to submit and write up topics that are intended to contribute to tomorrow's problems, not current ones. "We see the program as an option for the program manager to add to whatever his prime or R&D activity has to offer."

- **Topic Vetting.** The program executive officer reviews and evaluates all topics on which small firms might bid. Each program manager submits a prioritized list of topics, then competes in a rigorous process of SBIR topic selection and a vote is taken. SBIR contract awards are viewed as a reward, not a burden.

- **Treating SBIR as a Program.** This includes monitoring and follow-up of small businesses, not just making awards. He creates a spending plan for all SBIR projects by determining how many topics and contracts he can support within the PEO-SUB SBIR budget. He is vigilant in keeping an eye out for opportunities the small businesses can support with their technology and products. He encourages his program managers to demonstrate commitment by sharing the Phase II option costs; this amounts to $75,000, or half the cost. To make Phase III transitions easier, he has designed an acquisition planning process for program managers to award contracts to small business and clear a path toward follow-on awards.

- **Provide Acquisition Coverage.** He writes a broad Acquisition Plan that includes a list of all SBIR contracts let by his office and is available to all PEO-SUB program offices. Each year he adds information on the new awards including an approved plan by which every Phase II firm will seek to go on to a Phase III award.

- **Award Phase III Contracts.** This creates a convenient vehicle for program managers to reach small businesses, allowing single- and multiple-point solutions. The traditional burdensome DoD process has been streamlined for contracting convenience and flexibility. There is a contract ceiling of $75 million to allow local acquisition approval within the Program Executive Office.

- **Broker Successful SBIR Performers.** By pooling resources, money, and talent in one place, program managers can match successful Phase I/II companies with problems in program offices. This helps build the base of talent the office can draw on to solve problems for the submarine community.

- **Recycle Unused Phase I Awards.** Many companies never reach Phase II for a variety of reasons, leaving behind a potentially useful idea or technology. The Phase I database is a rich resource for the program managers who has a problem and needs an idea. Matching possible topics to the Phase I back-lists of reliable firms gives a manager a resource of "instant contracts" to solve a near-term problem. The topic or idea may go directly to a Phase II contract, which can reduce the topic cycle by 18 months and bring new technology to the fleet faster.

The PEO-SUB SBIR history includes more than 300 awards to 150 different companies since record keeping began in 1988. Since the first Phase III award in 1994, Team Submarine has made $1 billion in Phase III follow-on awards to about 15 companies. Three of the companies account for about $500 million. In summary, he said, "SBIR works." He showed a return-on-investment slide, indicating that a contribution of $126 million from Team Submarine ultimately spawned $1 billion of follow-on awards. Since all NAVSEA awards totaled about $1.5 billion dollars in the same period, the submarine program itself was very successful. He highlighted about 15 companies that made or soon will make the transition to Phase III funding, including Chesapeake Sciences Corp., Progeny Systems, and Digital System Resources.

Mr. McNamara concluded by encouraging managers to "take a good look at what you're doing, try to build in flexibility and follow the rules that Dr. Gansler drew up."

Stephen Lee
U.S. Army Research Office

Dr. Lee described himself as a "feet-on-the-ground" scientist who manages a research program that is focused on long-term, basic research. He writes topics, reviews proposals, oversees a process for Army reviewers to vet the topics, and designs strategy for his research organization and for the SBIR program. When writing a topic, he tries to anticipate who will support the Phase II step, Phase III and ultimately commercialization. Thus, because he worked more on the "push" side of the SBIR program than many participants, he suggested that his perspective would be different.

He noted that he could not place his activities on the highly complex DoD acquisition flow chart circulated at the meeting, describing his office as a "6.1 organization," which is the DoD's budget category for basic research. "Typically," he said, "we're feeding things before that chart begins." He said that the mission of the Army Research Office is to "seed scientific and far-reaching technological discoveries that enhance Army capabilities." This mission, he said, gave him a broad spectrum to investigate and enabled him to take some risks in the longer term. For this reason, he used the STTR program more actively than many people would—because of his basic research interest and involvement of academia.

An Emphasis on Research

The Army Research Office is divided into several directorates: engineering sciences, physical sciences, and mathematical and information sciences. Each directorate has three or four program managers who write SBIR and STTR topics. And while Dr. Lee said he has a good idea of what the Program Executive Officer for a certain subject area wants, he also has "a more fundamental, basic research, higher-risk thought pattern that I'm going through."

He showed an organization chart that illustrated how technology gathered from various sources (industry, academia, foreign laboratories, etc.) would typically transfer through SBIR programs in the direction of technology application (other services, program managers/program executive offices, Research Development Engineering Centers, other customers). This chart also showed that the Army Research Office was part of the Army Research Laboratory. The Army Research Office has no laboratories: it is an extramural funding office that supports research in universities and industries. While working with many customers outside the Army, including the Defense Threat Reduction Agency, the Army Research Office is responsible for "technology generation" and "technology enhancement" in the Army laboratory system.

Thinking about Commercialization from the Outset

Dr. Lee said that in writing topics, he tried to think from the beginning about where he would find support in moving it from Phase II to Phase III. He would also think about how to present a new proposal and move it toward a recommendation for funding. He conjectured that much of the success of his programs, like those of some other program managers in his position, is that they are "on the ground" and talking with the user, bringing the user directly to the small company and ensuring a connection from the beginning of Phase I when the topic is written. This is reinforced by the requirement to have a program executive officer or a program manager buy into the topic from the outset.

SBIR Projects that Benefited from Collaboration

He offered several examples of SBIR projects. The first was the Agentase Traffic Light Sensor, an STTR award with a co-Principal Investigator at the University of Pittsburgh. It was a defense technology of a type not discussed previously—a Homeland Security-based idea that was really dual-use. This one was a simple color-changing, enzyme-based assay incorporated into a new format with advantages over existing systems. It stabilizes enzymes in a polymer system. The sensor begins as yellow when wiped on a surface, then, if a nerve agent is present, it changes in less than two minutes to red. In the absence of a nerve agent it turns light green after about 15 minutes. The users worked with the developers to ensure that it was the right size, what it should look like, and whether it could be packaged in such a way that emergency personnel in safety suits could use it. As program manager, it was it was Dr. Lee's job to ensure such features of usability even before the chemistry was finalized.

Key to developing this sensor was the transition phase. Who should guide this process? The answer was the special operations man, who, in his responsibility for combating nerve agents, has a powerful interest in doing surface wiping. Interestingly, in the final commercial product, three colors proved to be too complicated and to bring potential liability issues. So the sensor begins at green and remains green if the environment is safe, turning red if contaminated. Some Phase III funding came from the Defense Threat Reduction Agency to ensure that the system would work with mustard agents and with blood and blister agents, as well as nerve agents, for deployment in Iraq. The process required close collaboration between Dr. Lee and the company in identifying the users. The largest sales for the company are now made to first responders.

The second example he cited was a solution called FAST-ACT, made by NanoScale Materials of Manhattan, Kansas, and based on research performed at Kansas State University. These were non-toxic reactive nanomaterials effective against a wide range of toxic chemicals, including chemical warfare agents. Dr. Lee wrote up an SBIR topic based in this area, and over several years the process had been scaled up and moved to Phase III, with the goal of producing multi-ton quantities of material. The material also required a great deal of animal testing, for which the Army Research Office had to identify one or more partners equipped to do this expensive work.

This product was now commercially available; it has been examined by the greater military, including the Special Forces, and used by first responders at Aberdeen Proving Ground as protection against chemical accidents. It was also being considered by the larger acquisition process and has been featured for sale by the influential Fisher Scientific.

Dr. Lee concluded by saying that he was particularly pleased with both of these SBIR projects, partly because of the close and successful collaboration between his office, the small company, and the users.

Tracy Van Zuiden
U.S. Air Force

Major Van Zuiden said that he has worked with the SBIR program for barely a year, after considerable experience in maintenance and logistics, and that he enjoyed hearing the lessons of those with longer SBIR experience.

He began with the Joint Strike Fighter, saying that it was the Air Force's vision to build an advanced and affordable strike fighter for the next generation for world wide customers. He said that SBIR projects would play a key role in the "advanced and affordable part of that equation." In the multi-service collaboration on the Joint Strike Fighter, the Air Force had 138 Phase I contracts and 57 Phase II contracts (total contracts), while the Navy had 64 Phase I awards, 54 Phase II awards, and 13 Phase III active contracts.

He said that his major challenge was in managing all these topics and contracts. The Air Force and Navy combined were generating approximately 70 new topics a year, and he thanked the SBIR offices of both services for helping with the complex management task. He listed the topics under four headings: Air Systems, Air Vehicle, Autonomic Logistic, and Propulsion. Then he broke down those topics by Integrated Product Teams to illustrate the kinds of technology projects in the program. He said that the important feature of these projects, which were used by the Integrated Product Teams, was that they have appropriate scope and relevance to move along a transition path to the final platform, the aircraft.

Success from the Point of View of Integrated Product Teams

He said he has thought about what makes SBIR projects successful from the points of view of the Integrated Product Teams, a small business, and the prime contractor. He suggested these behaviors for the Integrated Product Teams:

- **Accurately Define the Problem.** From the Integrated Product Team's point of view (represented by a floor engineer who travels extensively and is under pressure to move a product out the door) the most important tool is an accurate definition of the problem. It had turned out to be more difficult than he anticipated to define a problem so that a potential SBIR contractor understands what the SBIR office needs.
- **Don't Dictate the Path to a Solution.** It was difficult to persuade engineers to listen to all the topics before proposing a predetermined solution to a problem.
- **Actively Engage with the SBIR Contractor.** The program manager must do this not only during Phase I, and several times during Phase II, but continuously, to make sure the project is on track.
- **Involve the Prime Contractor and Supplier.** Involve the suppliers lower in the process than the prime contractors in the SBIR process. This is where the technology will be incorporated.
- **Promote Clear Communication Between Program Office, SBIR Contractor, Prime Contractor, and Supplier.** Communication is essential to SBIR success.

Success from the Prime's Point of View

Second, from the prime contractor's point of view, the keys to success are slightly different:

- **The SBIR Project Must Meet a Technology Need.** The prime contractor must see a technology push to realize the need for the SBIR. Prime contractors often assume they have a good enough solution, so they don't see a better one when it emerges from the R&D stage.
- **The Project Must Make Business Sense.** It is not sufficient to have a great technology: Is there a good business case for developing it? Can it be repeated? Is it cost-effective?
- **Willingness of SBIR Contractor to Partner with a Prime.** This is the path that leads to a working platform, as it is with the JSF.
- **Are Other Sources of Technology Available.** The prime has to be willing to look at sources other than the ones that are in use or that are familiar.

Success from the Small Business Point of View

Third, the picture again looks different from the small business point of view:

- **Know Your Customer and Keep Them Informed.** This, he said was supremely important—"it should be in red, with fireworks coming out of it." In proposing a technology for the Joint Strike Fighter, for example, where "weight is king," there is no point in designing a part that is heavier than existing technology.
- **Be Technically Accurate But Not Overbearing.** He said that sometimes a small business will be technically overbearing, as though to emphasize their competence. This, he said, is unnecessary and possibly counterproductive; the Army already knows that most innovation comes from small firms.
- **The Prime Contractor/Supplier Can Be Your Best Friend.** Although it can be hard to develop a relationship with prime contractors, they can be the best allies in helping to transition new technologies.
- **Build a Better Mousetrap.** The customer is counting on the small business for innovation, not "just the same mousetrap painted a different color."
- **Build a Sound Business Case.** The small business has to help the prime contractors and subs do this; no one knows the product better than the small firm, and they need to explain it to the prime contractors and subs.

He described a product that has been a success story, in several ways. The Army needed increased ear protection to dampen noise on jet aircraft, and a small firm developed new hearing protection. What made this SBIR product successful was that it turned out to have wide dual-use applicability not only to the Air Force and Navy but also in commercial applications. There was strong support from both the Air Force and Navy, which tried the task models and made helpful suggestions for improvement, as did the prime contractor. Also, the SBIR contractors realized that the product needed some variations to suit different users; the person who works at an admin desk in a flight-line operation does not need the same level of protection as the person who works on the deck. So the firm offered various levels of protection.

Major Van Zuiden also reported on a process that has been a success story. To find one, he reviewed the Joint Strike Fighter program to see which Integrated Product Teams did an outstanding job of managing, and found it was the propulsion branch. One reason was its long history of organized S&T, during which it did an excellent job of integrating its SBIR topics with the technology maturation process. Further, it enjoyed very active support from engine manufacturers, which is essential in the transition phase. And finally, it designated a full-time person to manage and promote SBIR projects within that Integrated Product Team. He concluded that this management pattern, which developed engines so sophisticated that they need almost no attention from the pilot, could provide a model for other technologies, such as data sensor fusion.

In summary, he offered the following measures to make SBIR more effective:

- **Involve the Prime Contractors and Suppliers**—especially including the suppliers two and three levels down, who will understand how the technology must be used.
- **Maintain Adequate Staff to Run the Process.** The SBIR office moved to web-based tools to help manage the process, but this cannot replace "the person behind the database."
- **Maintain Flexibility in the Use of SBIR Awards.** The traditional awards are $100,000 for Phase I and $750,000 for Phase II. There are few projects that can be done for a million dollars, he said, especially on a fighter aircraft. Even at the low end, projects cost from $3 to 5 million. "So we have to look at flexibility in the way we can use our SBIR products."

Peter Hughes
NASA Goddard Space Flight Center

Mr. Hughes introduced himself as the acting chief technologist at Goddard Space Flight Center and said he would talk about his perspective on the use of the SBIR program as "an investment tool in the R&D program." For FY04, the SBIR budget for the agency as a whole was roughly $110 million. At Goddard the SBIR budget was $14 million and the STTR budget was $3.5 million. The SBIR constituted about 20 percent of his "investments," in this sense, the rest being either internal investments, IRAD (internal R&D) core capabilities, Director's Discretionary Fund (designed to support high-risk, high-payoff efforts), or external competitive awards. He said that about six years previously, NASA experienced a major change from the historical custom of directing money from NASA headquarters through the program offices, without much internal or external competition. Most projects, including his own, must now compete for that money and adapting the culture to this change will require additional time and effort to be effective.

He began with the mission of NASA, which, he said, has "widespread recognition and brand appeal:" To understand and protect our home planet, to explore the universe and search for life; and to inspire the next generation of explorers. One exciting aspect of working with NASA, he said, was the ease of engaging the public, or anyone who wanted to participate in NASA programs. For the most part, he experienced a positive cachet in being affiliated with NASA. Even so, it has been difficult to sustain some worthy R&D programs because of the shift instituted by the President to emphasize the "moon, Mars, and beyond," but the basic drivers behind the mission and vision programs has changed little.

NASA Goddard, in Greenbelt, Maryland, employs about 3,000 civil servants and about 5,800 contractors. Its main foci are science, primarily earth and space science; the development of measurement instrumentation; and creating the platforms to make measurements. He broke down the scientific research further into earth science; making measurements of earth and earth systems; and space science, trying to understand the structure and evolution of the universe and planetary systems. A new area of research is called Vision for Exploration Systems. Underpinning all the science is technology development.

He described the core competencies of NASA Goddard as:

- **Experimental and Theoretical Science** that drives the instrumentation design and science measurements
- **Sensors, Instruments and Associated Technology,** especially in optics and electro-optics, that are used to develop sensors and instruments.
- **End-to-End Mission Systems Engineering,** with the capability to perform or lead implementation of all mission systems and operate scientific spacecraft.
- **Advanced Flight and Ground Systems Development** for, at any given time, two missions under development in-house involving about 10 to 20 instruments.
- **Large-Scale Scientific Information Systems,** to process, archive, extract (mine), and distribute data from multiple spacecraft and instruments to the science community, both inside the gates of Goddard and outside.
- **Program and Project Management,** both for in-house and extramural projects. At present, he said, Goddard has 19 active flight projects and about 25 in formulation. It also managed about 36 orbiting "space assets," about a dozen of them managed from Goddard, including the Hubble Space Telescope. It is "our pride and joy," he said, "to develop these

space assets and get them into orbit quickly to deliver to the science community the data they so desire."

Goddard was then heavily engaged in managing the development of the James Webb Space Telescope, which is to succeed the Hubble telescope. This platform is a huge challenge, he said, with a six-meter mirror that will operate in an Earth-Sun L2 orbit, about a million miles farther from the sun than Earth. It will allow sensing deep into space using infrared instruments. Despite daunting challenges across the entire mission, Mr. Hughes said Goddard has made good progress because of internal investments and investments through the SBIR program.

"Passing the Baton" from Phase II to Phase III

In trying to characterize what is important for the transition from Phase II to Phase III, Mr. Hughes used the analogy of a pair of relay runners on a track, where the lead runner who is slowing down must pass the baton to the runner behind who is speeding up. The vital element is a smooth, well-timed hand-off at the precise moment when both are running at the same speed. Managing the "hand-off" from SBIR Phase II to Phase III, he said, requires the same degree of communication and teamwork. Instead of an abrupt termination of work at the end of Phase II, the two phases must run "at the same speed" for a while until the Phase III work moves out on its own.

NASA has recorded a number of visible successes in transitioning their technologies into major programs, he said. One was mentioned earlier by Carl Ray—the Mars Exploration Rovers, where Goddard embedded several technologies, including lithium ion batteries, paraffin-based heat switches, and customized commercial microchip technology on the 2003 Rover. Goddard also managed development of a number of technologies for the Aura mission,[38] including composite optics, a radiometer, and signal conversion chips. This was a good example of a NASA SBIR program, he said, because well before the start of the mission, Goddard listed the challenges it saw in these areas and was able to infuse some of the technologies as they matured by maintaining close coordination between the program management and the technology programs at Goddard.

NASA Goddard also tries to play a unique role in teaming and combining a number of component technologies, or system technologies that come from SBIR and other internal investments in unique ways. Some phenomena have a multi-

---

[38]The Aura (Latin for "breeze") mission researches the composition, chemistry, and dynamics of Earth's atmosphere, including studies of ozone levels, air pollution, and climate. Aura is part of the Earth Observing System (EOS), a program dedicated to monitoring the complex interactions that affect the globe using NASA satellites and data systems.

plier effect when combined, but when they were integrated, they sometimes produce unique measurement properties not anticipated at the outset.

What Works, What Doesn't Work

He then moved to procedures that are working and not working for the SBIR program. First, he said, it was his personal belief that NASA should use the SBIR program as a new and unique source of innovations. He viewed SBIR as "his ERAD," the External R&D program. He had considerable resources for IRAD (Internal R&D), but he wanted to be able to reach out to new R&D communities and try to bring them into the fold at Goddard. He wanted to "break the mindset" that the SBIR program was not just a set-aside for small business, but rather an external source of innovation that we want to include as members of Goddard's R&D team.

Second, he said that he wanted to push the subtopic managers to identify really tough problems that were tough to solve and put them out to small firms. He did not want to "just lob some softballs" to these firms; he wanted his managers to be "really pitching sliders" that were going to require real creativity to solve.

Third, he stressed the importance of maintaining clear strategic priorities in these technologies and inducing the Phase II and Phase III proposers to try new directions in their technologies where applicable.

Fourth, he said it was essential to have top talent review these proposals—people who understand the technology and who were going to be Contracting Officer's Technical Representatives (COTRs). In his early years at NASA, he said, he was amazed at how often these reviews were done by junior people when no one else was available. He said that some of the most successful SBIR elements had the very senior, technical, experienced reviewers and the COTRs providing oversight.

Fifth, he emphasized the importance of looking for and helping the SBIR Principal Investigators develop a realistic work plan that could be accomplished within the stated period of time. Often the small firm is "so enthralled with what they propose to do" that the program manager does not ensure that they have thought through the development process thoroughly. SBIR firms often encounter a culture shock in working with the agency and the rigorous systems engineering to which NASA technology is subjected before it can be put into operation. A Principal Investigator new to the NASA system must understand that space systems have one and only one chance for success and that there is no way to correct for failure after launch. For that reason, the program managers want to avoid any risk whatsoever, and need to work with the SBIR firms on a clear and specific risk-mitigation plan.

Finally, he said that NASA must consider carefully the long incubation process for new technologies, which may take five, six, or seven years. This must be anticipated in the planning from the outset in order to anticipate whether the project can really by ready in time for infusion into the larger system.

Factors that Bring Success

He then outlined some of the success factors for execution that NASA had discovered over the years:

- Keep the SBIR program managers and the subtopic managers focused on the strategic priorities for the technology and redirect them when appropriate.
- Encourage the task manager/subtopic manager to work closely and interactively with the Principal Investigators. It is not enough for them just to agree on the parameters of the grant and then meet at the conclusion of the program. Those Principal Investigators who work closely with the topic managers throughout development have the most success.
- Encourage the subtopic manager and Principal Investigator to actively engage in trying to identify infusion opportunities. This involves close teamwork with the NASA office of technology transfer to understand the market potential both inside Goddard and at other agencies.
- Leverage the cachet of working with NASA. At the same time, the company must independently try to take the technology all the way to an infusion point with NASA and understand the rigors of doing so.
- Keep these SBIR tasks off the critical path. They should not be used for development until the risks are understood and a strategy for reducing those risks has been set.

Mr. Hughes said that what is working well in the SBIR program was the special procurement authority available at Phase III. If projects had to be reopened to general competition at this stage it would not work.

Where to Add, Where to Strengthen

He summarized the procedures that should be added or strengthened, as follows:

- Increase the focus by SBIR technical managers on the strategic value and utilization of the technology—not solely on completing a Phase III contract.
- Bring more rigor to the analysis of the Phase III return on investment and the factors that contribute to strong returns.
- Find some way to overcome the poor access to Progress and Final Reports of SBIR projects at other agencies and even other NASA centers, in order

to take advantage of technologies that are dual-use or applicable to other programs.

- Add a mechanism that permits some mid-year start-ups to meet new, urgent needs. Such a mechanism was needed when the new Vision for Space Exploration was launched the previous year.

DISCUSSION

Mr. Levine, the moderator, asked other members of the panel to suggest their own improvements for the SBIR program.

Mr. McNamara said that for him the program was running well. He pointed to the lack of pre-planned budget for SBIR Phase III transitions' Valley of Death as a general concern, and suggested that higher dollar values for Phase II would help in the transition to Phase III. He also urged more education of people on how best to use the program. Other than that, he praised the acquisition tools and the sole source follow-on as making the program powerful.

Mr. Levine agreed that major strengths of the program are the flexibility of Phase III procedures, especially the ability to start a Phase III project with few bureaucratic hurdles. He suggested that more people, even within the military, would be interested in knowing more about these advantages. He also said the program should be improved by speeding up the process of producing and publicizing the topics. In terms of the Valley of Death, he called for a better mechanism to help firms doing fundamental science plan how to scale up or build a prototype after the principle has been proved.

Major Van Zuiden said he would like "to steal an idea from NAVAIR," which was a concept called "clustering" of SBIR awards. This might occur for problems that proved too complex for one or two SBIR projects and might require that five, six, or seven firms work together on a common problem, each taking a different aspect. When that project is finished, a prime or perhaps an SBIR contractor might integrate the work to form a solution that can be commercialized. He added that he would like a better way to "spring-load the Phase III"—perhaps by placing interim milestones in Phase II that would be linked with some bridge funding before commitment to the full terms of the Phase III.

Dr. Wessner said that the suggestions all sounded helpful, and asked why they could not be implemented. Mr. McNamara said that when he needed a Phase II award that was larger than the official limit of $750,000, he often received a flexible response from the Navy program manager, John Williams, with extra funding to help reach Phase III. "So when you want to pull the trigger and go to Phase III," he said, "I think the rules basically allow you enough flexibility."

Navy-Air Force Collaboration

A questioner asked whether anything should be done to improve the collaboration between the technical communities in the Navy, especially the submarine program, and the Air Force. Mr. McNamara said the two communities did run "in parallel," and were moving closer together. The submarine technology effort focuses its R&D efforts on mainstream R&D research and on transitions from 6.2 (applied research) to 6.3 and 6.4 (development and applications). That is usually focused through the Navy laboratories, he said, producing a buy-in and a process to follow. When a "disruptive" (innovative) technology becomes available, it may not draw immediate interest in this more traditional environment. SBIR awards allow the effort to by-pass that environment and to jump-start the new technology or allow it to compete for attention. He said that the SBIR efforts were complementary to the Navy laboratories and could accelerate transitions from Navy activities into tactical systems.

Major Van Zuiden said that for submarines, the technical people in the laboratories helped evaluate SBIR projects. He added that as some of the laboratory budgets decrease, with reductions in defense spending, they will probably use SBIR awards to augment their own research and development.

Dealing with the Problem of Timing

A participant asked Mr. McNamara about his comment that the SBIR awards should be used for tomorrow's problems, not today's. Part of the problem with the transition, however, comes when the results from Phase II research emerge after three or four years only to find that the original requirement had disappeared or that the new technology was too disruptive to be introduced smoothly. He also referred to Mr. Hughes' comment that the timing of the infusion is very important, and if SBIR development is out of the R&D mainstream there might already exist other solutions.

Mr. McNamara agreed that by the time a topic is written, vetted, and advertised, and a small business is selected for an award, the time has passed to solve "today's problem." If he had today's problem to solve, he said, and he wanted to use an SBIR, he would look for a Phase I result to recycle. This would give him a head start. He might also look at a number of sources, whether a prime contractor or a Phase III SBIR, that are already working on the question or a related question, or look at a laboratory.

Does Advocacy Bring Success?

The questioner said that Mr. McNamara's seemed to be a very successful SBIR organization, with by far the largest number of Phase III awards. He noted that as executive director, Mr. McNamara devoted much of his time to advocating for the SBIR program. He asked if this advocacy was the essential role of the program manager, to promote the introduction of science and technology into acquisition programs, whether this would be appropriate for other science and technology activities. He also asked whether the other science and technology areas were transitioning to his program as he had SBIR transitioning to his program.

Mr. McNamara said he was active as an SBIR advocate because before he arrived, the SBIR had been largely overlooked or assigned low value, with no linkage to acquisition. He said that the acquisition offices were the key agents in moving SBIR projects into Phase III. Also, he said that his office now did "a pretty good job," partly because he had been around long enough that the SBIR was ingrained in the organization. People now wanted to work with it and welcomed opportunities to submit their topics.

He added that some of his sister organizations did not seem to be quite as interested yet. They did use the SBIR money and had started to treat it like a program, but they had not had Phase III successes. He said that a common problem for them was that they could not interest the prime contractors in their projects. The submarine program tried to get its SBIR projects to the prime for each new technology under development. It also had an instant market in the in-service community, which allowed them to move their products out to other markets within the Navy much more rapidly. The best strategy, he said, was not to tie a project to a single large program, such as a single Virginia-class submarine, but to introduce several dozen new technologies or "back-fits" per year, and to become known for producing specified R&D solutions.

Lessons Learned

Moderator:
Jacques S. Gansler
University of Maryland

Dr. Gansler said that the last question of the day was both important and difficult: To improve the program, where is the best place to focus attention: on the program office? the firm itself? the prime contractor? on the policy or Congressional levels? He said that the last panel had representation for each of the four areas that had been discussed throughout the day: Congress, the prime contractors, the services, and the small business community. He asked representatives of each for key points they would take away from the conference.

Richard Carroll
Innovative Defense Strategies

Mr. Carroll said he had been involved in the SBIR program since its inception, and offered a quick overview of his experience. He started a small business in 1982, and competed on a number of SBIR programs. The company built up an SBIR inventory of Phase I and Phase II awards, and developed a large number of Phase III projects. He had sold the company to General Dynamics about 19 months previously, when it had significant Phase III awards, which the new owner was using to compete in the marketplace.

He said he was also closely involved with the reauthorization of the SBIR program in 1992. It was during that reauthorization that many Phase III issues were debated and the importance of Phase III became apparent, so the Congress added significantly to that phase of the program. He worked with the Small Business Administration, the DoD, and the Office of Management and Budget in helping to frame new Phase III initiatives for the new SBIR directive issued in 2000. When the directive was issued, it became clear that Phase III was a viable pathway for SBIR businesses, resulting in many Phase III activities. The bill also called for the current study to evaluate the program, compare it with other science and technology initiatives, and recommend improvements.

How to Fund Phase III

He said that he was still working to improve Phase III and involved in debates on how to fund Phase III activities, which "remains one of the most significant issues in the program." There was discussion about the Senate and the House Armed Services bill in 2005 that dealt with the subject. He brought to the conference several charts to illustrate the process of producing defense contracts, including purchasing and acquisition pathways, as well as other science and technology activities throughout the agency. The charts were extraordinarily complex, with many dozens of acronyms. And yet, he said, in order to make Phase III successful, the SBIR program had to be able to incorporate or feed into that labyrinthine process "in a way that makes sense to program managers." He said that he believed it could be done, however, in a meaningful way so as to avoid the Valley of Death.

The Advent of Spiral Development

Dr. Gansler added that a major change now taking place in the acquisition process is "spiral development,"[39] a method that is designed to constantly increase capability at lower and lower cost. This process brings an opportunity to take full advantage of the SBIR program, he said, by introducing in each subsequent cycle the changes that have been made in the previous cycles. The next challenge that has been discussed in this conference is to apply appropriate testing and evaluation to reduce risk to an acceptable level. This technique, which is being used for the Joint Strike Fighter, is an excellent focus for the SBIR program—to demonstrate techniques that lower risk and introduce them in subsequent cycles.

---

[39]Spiral development was introduced in the mid-1980s as a way to reduce risk on large software projects. A spiral, or cyclical, approach is one that allows customers to evaluate early results and in-house engineers to identify potential trouble spots at an early stage. On subsequent "turns" of the spiral, early changes are incorporated, additional evaluation is done and changes made. The Department of Defense has adapted the technique as part of its evolutionary acquisition strategy to move newer technologies into large platforms, such as assault vehicles and computer systems, much more quickly. <*http://www.washingtontechnology.com/news/18_5/cover-stories/20881-1.html*>.

John P. Waszczak
Raytheon Company

Mr. Waszczak said that his summary comment would begin with the cultural barriers that must be overcome to be successful in Phase III. A critical element for doing this is technology roadmaps that are integrated by government, prime contractors, sub-prime contractors, and small businesses. Another is program manager and Program Executive Officer pull, as well as a plan to transition from Phase II to Phase III before beginning Phase II. This means that a funding pool and insertion plan is needed for each projects.

He also extolled the matchmaking effort that John Williams and his team from the Navy and others had developed to match prime contractors with small businesses. This is expedited by the Navy's Technology Assistance Program and similar gatherings by the Defense Department, NASA, and others.

The Advantage of Beginning with "Phase Zero"

To overcome small businesses' difficulty in selling to prime contractors, he said that the key element is to start working with small businesses in "Phase Zero," before Phase I even begins, so that they can better prepare for insertion as they go through Phases I and II, building trust with the customer and the prime, and trying to mitigate the perceived risks of working with a small firm.

He underlined the importance of incentives, and said that what truly motivates prime contractors—and small businesses—is return on investment. Because prime contractors, like every other organization, are resource constrained, they need to make sure the Phase III process is efficient if they are going to invest in it. They must also be sure that the customer and the end user—the warfighter—receive the benefits.

Finally, he said, the prime contractors needed good metrics to evaluate the Phase III transition and the performances of customers, prime contractors, and small businesses. This conference, he said, had produced many good suggestions for metrics and had strengthened the network for sharing them across the different services, agencies, and prime contractors.

Dr. Gansler interjected that the three large prime contractors at the conference had all indicated that within the past year, they had noticed a shift toward more pull from the prime contractors, which is one of the key elements in making this program a success.

John Williams
U.S. Navy

Mr. Williams said he had always been a strong proponent of Phase III transitions, and of keeping the focus on Phase III throughout the SBIR program. Phase

III transitions involve multiple partners who must all address the same issues together, as they had been doing at this conference.

Noting that the Navy has a technology-pull approach, while the other services tend toward a technology-push approach, he suggested that it was always more challenging to transition technologies from a research or laboratory environment into an acquisition program. The Navy SBIR program has evolved to a point where the people who control the money for SBIR awards—the Program Executive Offices—are the people who plan the Phase III transition; and, their acquisition program offices directly influence the flow of Phase III dollars.

He recommended that the most important metric should be Phase III funding to the SBIR firm, and that those SBIR-managing Navy offices that generate significantly more Phase III follow-on funds should be rewarded. The Navy program, he said, basically gives back the 2.5 percent SBIR assessment to all assessed Program Executive Offices, but he feels that providing more funds to those Program Executive Offices that generate more Phase III metrics makes better sense. Since 2002, the Navy has awarded more Phase III dollars (non-SBIR funds) than Phase I and II SBIR dollars combined. In fact, for every SBIR dollar, Navy SBIR projects raise almost two dollars in additional funds. He said that this was one result that Congress likes to see, and would like to see more.

Using the Company Commercialization Report

He also said that SBIR firms should be able to compete for the "other" 97.5 percent of extramural RDT&E dollars, in addition to the 2.5 percent reserved for SBIR. Accordingly, the Navy uses a Company Commercialization Report (CCR) through which SBIR and non-SBIR revenues for small firms are identified. Even for small firms winning many tens of SBIR awards, he said, SBIR funds represented about 25 percent of their revenues—never more than 50 percent. That is a good sign, he said; if a company continued to win only SBIR awards, but not other awards, it's likely that its technologies are not being transitioned. He recommended that the Academies use the CCR for measuring Phase III results of Phase II awards. The CCR captures not only Navy SBIR firms, but also many small firms working with the other agencies.

One challenge in building the Company Commercialization Report is that those in government who are evaluating the data are typically engineers who lack experience in business. He recommended that the NRC committee consult with leading economists and other experts to develop insight about how to gather the most relevant data and how best to evaluate them to measure effectiveness in commercialization.

Blending a New Technology with Other Technologies

He mentioned a business technical assistance program first initiated for the Department of Energy, and to some extent for NIH and NSF, by Dawnbreaker, Inc., that educates firms about how to commercialize technologies. In the Navy version of this program, he said, the objective is to show small firms actual requirements and reports describing the specific needs of the customer. This would include the details of other technologies it would have to work with, whether it needed to be licensed, and other information.

A paradox for the Navy SBIR program is that it has been so successful with their Transition Assistance Program program that it is beginning to drain its administration budget, given that it costs the Navy about $15,000 per small business participant. In one sense, this does not appear to be a serious expense—within 20 months of the program, participating firms average $3 million in Phase III awards, so the return on investment is almost a hundred-to-one. Nevertheless, he noted that it is still difficult to find additional dollars to put on a SBIR project to encourage commercialization because of Title 301 in the 1992 legislation, which describes what funds could be used for technical assistance. Because of the way this rule is written, he said, it is impossible to conclude whether the same percentage of funds allowed to support technical assistance in Phase I can be used in Phase II. He requested clarification of the rule, since the ability to use such funds for Phase II would provide up to $12 thousand per small firm to SBIR agencies to help SBIR companies.

DISCUSSION

James Rudd of the NSF said that his agency was involved in commercialization, and asked the DoD participants whether they would declare their SBIR program a success if a certain percentage of their R&D funding came from the SBIR program. Mr. McNamara responded that for an organization such as his, which spent between $5 and 6 billion a year, much of it on submarine procurement and costly ship maintenance, the SBIR money was a small amount, but when used, it has a great return on investment. Nonetheless, in parts of his organization, especially in electronics, where R&D is central, SBIR played a larger and more important role.

The Cost-effectiveness of Small Business

Another questioner asked how the services would rate the impact of SBIR companies on their overall innovation and business activities.

John Williams replied that the impact was large. He added that small businesses, especially those coming through SBIR, were about 30 percent cheaper than large businesses, and also cheaper than laboratories and field activities. In addition, he noted that some small firms had proven more reliable than large firms. More broadly, he said, the process benefited by "adding people to the landscape" and producing more options for the services. Moreover, a lot more innovation had occurred at the big firms, he said, because they knew they were being measured against organizations other than their peers.

Performance Improvements at Lower Cost

Dr. Gansler added that a lot of SBIR activity occurred not at the weapons system level, but at the sub-system, component, or even the materials levels. For example, SBIR contributions often brought performance improvements at lower cost. This, he said, is "the way the real world, the commercial world, actually operates," and that government should try to follow the lead of the computer industry, which produces new models at lower cost every two or three years.

A New Metric to Gauge Phase III Success . . .

Another participant said that dollar amounts and percentages were less valuable as metrics than the results of the Phase III contracts as requested every year by the Small Business Administration (SBA). "That," he said, "is the easiest and cleanest metric" to indicate success, and it would be inexpensive. The SBA metrics would not provide a complete picture, he said, but they could show trends. In the same way, the achievements of the prime contractors could be measured. Success stories were not helpful, he suggested; every large program should have many success stories; and there were no clear metrics by which to measure innovation. The program should focus simply on measuring how much additional money was going to SBIR firms. "That was the intent of Congress," he concluded, "and an easy way to measure it."

. . . and a Caution about this Metric

Larry James of the Department of Energy urged caution about using the simple metric of Phase III dollars, which might drive the program further toward the development side. The "Innovation" in SBIR would eventually be lost. The program will gravitate toward low risk, incremental development projects. The program's major benefit to the nation is that Small Business Concerns are willing and able to take on high-risk, innovative research that provides disruptive technological advances in the marketplace.

Concluding Remarks

Jacques S. Gansler
University of Maryland

In calling the conference to a close, Dr. Gansler noted that the broader goal of the SBIR program is for the results of research to be utilized. The idea was not to emphasize Phase III at the expense of Phases I and II, but to do it in addition to Phases I and II. Every science and technology organization faced the same challenge of finding the right balance between the "R" and the "D."

In fact, he concluded, research needs vigorous emphasis at every level—certainly at the level that produces revolutionary ways of doing things—if we as a nation are going to maintain our leadership in technological innovation and economic development.

III

APPENDIXES

Appendix A:
Biographies of Speakers*

MICHAEL CACCUITTO

Michael Caccuitto serves as the SBIR/STTR Program Administrator for the Department of Defense (DoD) within the Office of Small and Disadvantaged Business Utilization (SADBU) in the Office of the Secretary of Defense (OSD). In this role, he is responsible for policy implementation and program administration across DoD, while program execution and management occurs at the component level.

Prior to joining the OSD SADBU staff in February 2005, Mr. Caccuitto served on the staff of the Deputy Under Secretary of Defense for Industrial Policy. He worked a variety of industrial base issues during his five years in Industrial Policy focusing particularly on transformation, innovation, and emerging supplier access to DoD.

Before joining the OSD staff in February 2000, Mr. Caccuitto served for nine years in the U.S. Air Force in a variety of research and development, acquisition program management, and staff roles, in active duty and reserve capacities. He remains in the Air Force Reserve assigned to the Assistant Secretary of the Air Force for Acquisition.

He holds master's degrees from the Harvard University's John F. Kennedy School of Government and Rensselaer Polytechnic Institute, and a bachelor's degree from the University of Rochester.

---

*As of June 2005.

RICHARD CARROLL

Richard Carroll founded Digital System Resources, Inc. (DSR), a system integration and software company specializing in technology critical to national security. The company was formed in 1982, incorporated in 1985, and has grown to 480 people with net revenues for 2003 of over $125 million. DSR now is in the top 100 largest prime Department of Defense contractors for Research, Development, Test, and Evaluation, and is a recognized leader in providing state-of-the-art, high-quality products.

DSR, under the leadership of Richard Carroll, has taken on the challenge of introducing a new software model to defense systems. DSR's products and services have been recognized with numerous awards and a continuum of competitive contract awards. DSR's experience includes the development and production of systems for passive and active sonar, electronic warfare, combat control, and computer-based training and simulation for these systems. DSR has an outstanding record of delivering these systems on time and within budget.

Richard Carroll has been called upon on several occasions to testify on the role of small high-tech business in providing innovation. He has become a recognized expert on the potential of small high-tech businesses to provide cost-effective solutions to complex problems. In particular, he has testified on the importance and limitations of the Small Business Innovation Research program in meeting the need for government innovation.

In 2003, DSR was acquired by General Dynamics Advanced Information Systems (GD-AIS). GD-AIS brings significant opportunities for the rapid use of DSR-developed technologies throughout the defense marketplace.

In an effort to promote small business success, Richard Carroll joined the new Small Business Technology Coalition (SBTC), became a charter member, member of the Board of Directors, and served as its chairman between 1999 and 2001. Richard Carroll works closely with the Legislative Committee of SBTC and was responsible for getting SBIR reauthorization through Congress. Through the activity of the Legislative Committee, SBTC has become the recognized small business authority on SBIR reauthorization within the Small Business Administration and Congress.

Richard Carroll serves on the board of directors of the Naval Submarine League. The League has the mission of promoting the advancement and a better understanding and appreciation of the need for a strong United States submarine fleet. He was also a member of the panel making recommendations to the Governor of Virginia on technology issues in the state and on the advocacy of small business contributions to employment and business growth in the Commonwealth of Virginia. Mr. Carroll was selected as the 1999 Ernst & Young Master Entrepreneur of the Year for the mid-Atlantic region. He was also a finalist for the Ernst & Young Entrepreneur of the Year in 1998 and was selected by the Virginia Secretary of Technology as the Virginia Entrepreneur of the Year for 1999.

DSR and Lockheed Martin received a joint Hammer Award from Vice President Gore's National Partnership for Reinventing Government. This reinventing government award was for DSR and Lockheed Martin's collaboration on the Acoustic Rapid COTS Insertion (A-RCI) program which regained acoustic superiority for the U.S. submarine fleet.

Mr. Carroll received a B.S. in physics, University of Vermont, 1976; and did graduate studies at the George Washington University, 1977-1978; and the University of Texas, 1978-1979.

TOM CASSIN

Tom Cassin is the president of Materials Sciences Corporation. Mr. Cassin has over 15 years' experience in the analysis, design, and fabrication of composite material structures. Mr. Cassin is responsible for the development and execution of technology programs for the application of composite materials. Since joining MSC in 1987, Mr. Cassin has been involved in several different aspects of composite material applications. In the area of micromechanics, Mr. Cassin has been responsible for the development and application of physically based materials characterizations in brittle and ductile failure applications. Mr. Cassin has authored several of MSC's in-house analysis codes and has presented work in these areas in scientific journals and papers. For the last decade, Mr. Cassin has been primarily involved with use of composites for Naval applications. He has acted as program manager on several applications of composites with the most recent work involving design, fabrication, and evaluation of solid and cored fiberglass materials for topside Naval structures.

Mr. Cassin has B.S,M.E and M.S.M.E degrees from Villanova University and is a member of ASME, AIAA, SAMPE, and Pi Tau Sigma.

THOMAS CRABB

Thomas Crabb is vice president, chief financial officer, and treasurer of Orbitec. Headquartered in Madison, ORBITEC is Wisconsin's aerospace research and product development leader, proving very strong in the use of the Small Business Innovative Research program as a catalyst for technology and product development. ORBITEC has had over 130 government contracts exceeding a total of nearly $92 million.

Mr. Crabb provides technical, managerial, entrepreneurial, and financial capabilities and experiences enabling a unique perspective for strategies, operations, and developments for business. He has led major areas of business management including finance, accounting, large program development, marketing, proposal preparation, market research and business plan developments, patent development, quality, information technology, commercial product development, and project and personnel management. Mr. Crabb spearheaded, with two senior partners, ORBITEC as a high-tech aerospace company headquartered in Madison, Wisconsin. To continue his vision for technology expansion into commercial markets, he developed an approach and incorporated the sister company, Planet Products Corporation (PLANET), which later transformed into PLANET LLC. He has been the key to ORBITEC's expansion and success.

As leader of ORBITEC's largest profit center (of two), Mr. Crabb is involved in the strategic development of ORBITEC's corporate technical, financial, and operational assets. He has been responsible for ORBITEC's major and sustained growth, which include areas of spaceflight hardware development, environmental chambers and control systems, life support systems, sensors and instrumentation, 3D training systems, commercial product development. He is responsible for ORBITEC's first patents, first commercial product, and first spaceflight hardware program. Mr. Crabb is directly responsible for attaining NASA's two largest Phase III SBIR awards for flight programs that were led by Mr. Crabb at $57 million and $33 million. He also led the establishment of ISO certification, awarded in January 2005.

Mr. Crabb's current professional activities include Board of the Small Business Technology Coalition; member of the American Society of Gravitational Space Biology; senior member of the American Institute for Aeronautics and Astronautics (AIAA); and member of the Wisconsin Space Grant Consortium. Previously Mr. Crabb was chairman of the AIAA Space Systems Technical Committee (STTC); secretary of the Space Operations Technical Committee, member of the National Space Society; member of the Planetary Society; chairman of the AIAA Wisconsin Chapter; proposing founder of the Wisconsin Center for Space Automation and Robotics; and past director of Outreach–Wisconsin Space Grant Consortium. Mr. Crabb has also taught space system design coursework at the University of Wisconsin, Milwaukee.

Over 60 reports, publications, and contributions to a wide range of technical areas are available on request. He was a finalist for 2002 Entrepreneur of the Year. Other awards include Superior Performance Award, NASA Johnson Space Center; First Shuttle Flight Achievement Award; Governor Thompson Award for High Tech Day, 1987; Wisconsin Innovation and Research Awards (each year 1988-2000); ORBITEC 1996 SBA/SBIR Tibbetts Award for Small Business; ORBITEC 1999 SBA/SBIR Tibbetts Award for Small Business; and Madison Civics Club Recognition for Outstanding Service (2001).

Mr. Crabb has earned the following educational degrees: B.S., engineering mechanics and astronautics, The University of Wisconsin, Madison; M.S. engineering mechanics—Aerospace Option, The University of Wisconsin, Madison (began M.S. at the Aeronautical and Astronautical Engineering Department, The Ohio State University); M.B.A. at The University of Wisconsin, Madison.

He has also taken the following training courses and classes: Independent finance and accounting coursework, University of Wisconsin, Madison; Large

Program Management; Technical Writing; Program Management Systems; Proposal Preparation; other courses relating to management and business.

JACQUES S. GANSLER

The Honorable Jacques S. Gansler, former Under Secretary of Defense for Acquisition, Technology and Logistics, is the first holder of the Roger C. Lipitz Chair in Public Policy and Private Enterprise at the University of Maryland, College Park. Dr. Gansler is also the chair of the National Academies SBIR study committee.

As the third ranking civilian at the Pentagon from 1997 to 2001, Dr. Gansler was responsible for all research and development, acquisition reform, logistics, advanced technology, environmental security, defense industry, and numerous other security programs. Before joining the Clinton Administration, Dr. Gansler held a variety of positions in government and the private sector, including Deputy Assistant Secretary of Defense (Material Acquisition), assistant director of Defense Research and Engineering (Electronics), vice president of ITT, and engineering and management positions with Singer and Raytheon corporations. Throughout his career, Dr. Gansler has written, published and taught on subjects related to his work. He is the author of *Defense Conversion: Transforming the Arsenal of Democracy*, MIT Press, 1995; *Affording Defense*, MIT Press, 1989, and *The Defense Industry*, MIT Press, 1980. He has published numerous articles in *Foreign Affairs, Harvard Business Review, International Security, Public Affairs,* and other journals as well as newspapers and frequent Congressional testimonies. He is a member of the National Academy of Engineering and a Fellow of the National Academy of Public Administration.

BILL GREENWALT

Mr. Bill Greenwalt joined the staff of the Senate Armed Services Committee (Senator John Warner, Chairman) in March 1999 and is responsible for defense acquisition policy, information management, industrial base, export control, and management reform issues. He is also lead staff member for the Subcommittee on Readiness and Management Support. Previously, he served on the Senate Governmental Affairs Committee (Senator Fred Thompson, Chairman) as a professional staff member responsible for federal management issues and committee press relations.

Mr. Greenwalt served as a staff member for the Senate Subcommittee on Oversight of Government Management and as a military legislative assistant to Senator William Cohen, where he was responsible for legislative efforts to reform federal information technology acquisition, culminating in the Clinger-Cohen Act of 1996. Prior to coming to the Senate in 1994, Mr. Greenwalt was a visiting fellow at the Center for Defense Economics, University of York, England, where he served as a country expert on several studies for the European Commission. Previously, he worked for the Immigration and Naturalization Service in Frankfurt, Germany, and also as an evaluator with the U.S. General Accounting Office in Los Angeles, California, where he specialized in defense acquisition issues.

Mr. Greenwalt graduated from California State University at Long Beach in 1982 with a degree in economics and political science and received his M.A. in defense and security studies from the University of Southern California in 1989.

RICHARD H. HENDEL

Richard Hendel is a principal specialist in the Enterprise Supplier Diversity Program Office at The Boeing Company. He is located in St. Louis. Mr. Hendel is the Small Business Innovation Research program manager for The Boeing Company. As such he has the responsibility for developing and implementing strategies that will expand utilization of the Small Business Innovation Research program, the companies and their technologies and products across the Boeing Enterprise (Businesses and Functions). To accomplish this he works with various programs and personnel in the Boeing Phantom Works Engineering & Information Technology and Structural Technologies organizations and the Boeing Integrated Defense Systems business unit, as well as small businesses, concerning SBIR projects and activities. He represents the company at national SBIR conferences and has made presentations on the subject at national, regional, and local conferences.

Mr. Hendel has been with Boeing for 28 years and held various positions in the Subcontract Management and Procurement organization prior to joining the Supplier Diversity Program in 1989. He received MDA Teammate of Distinction awards in 1994 and 1995; and an MDA Leadership Award in 1996 for his contributions to the, at the time, McDonnell Douglas small business program. Additionally, Mr. Hendel serves on the Incubator Advisory Committee of the St. Charles County Economic Development Council. He served as the 1998 St. Louis Small Business Week Committee chairperson and continues to serve on this annual event's planning committee.

Mr. Hendel received his undergraduate degree from the University of Missouri, Columbia and his graduate degree from Webster University in St. Louis. Rich is a member of the American Legion, the Knights of Columbus, and is a member of the American Baseball Coaches Association.

CHARLES J. HOLLAND

Dr. Holland is the Deputy Under Secretary of Defense (Science and Technology). He is responsible for Defense Science and Technology strategic planning, budget allocation, and program review and execution. He ensures that the National Defense objectives are met by the $9 billion-per-year DoD Science and Technology program. Dr. Holland is the principal U.S. representative to the Technical Cooperation Program between Australia, Canada, New Zealand, the United Kingdom, and the United States. He is also responsible for the DoD High Performance Computing Modernization Program, the Defense Modeling and Simulation Office, and management oversight of the Software Engineering Institute.

Previously, he was Director for Information Systems within the ODUSD(S&T). He formulated guidance, developed the strategic plans, and provided the technical leadership for the entire DoD information technologies R&D effort, with an annual budget of approximately $1.8 billion. Technology programs under his purview included decisionmaking; modeling and simulation; high performance computing; information management, distribution, and security; seamless communications; and computing and software technology. He served as the DoD representative to the interagency Critical Infrastructure Protection R&D group responding to Presidential Decision Directive 63.

Prior to being appointed the Director for Information Technologies in March 1998, Dr. Holland was the Director of the DoD High Performance Computing Modernization Program Office reporting to DUSD(S&T). A substantial portion of Dr. Holland's government career involved the direction of basic research programs in applied mathematics and information technology at the Air Force Office of Scientific Research (1988-1997) and at the Office of Naval Research (1981-1988). He served as a liaison scientist at the European Office of Naval Research in London from 1984-1985.

Prior to joining government service, Dr. Holland was a faculty member and researcher at Purdue University and the Courant Institute of Mathematical Sciences at New York University. He has authored more than 20 research publications on control and systems theory, probabilistic methods in partial differential equations, and in reaction-diffusion phenomena. He is professionally recognized, along with his co-author, Dr. Jim Berryman, for the analysis of fast diffusion phenomena.

Dr. Holland was an Army ROTC graduate in 1968. Following an education delay for graduate school, he served as a 1st Lt. in the U.S. Army, Military Intelligence, in 1972.

Dr. Holland received the Presidential Rank Award, Meritorious Executive (2000) and the Society for Industrial and Applied Mathematics Commendation for Public Service Award (1999). He is a recipient of the Meritorious Civilian Service Award from the Secretary of Defense (2001), Air Force (1998), and the Navy (1984).

Dr. Holland received a B.S. (1968) and an M.S. (1969) in applied mathematics from the Georgia Institute of Technology and a Ph.D. (1972) in applied mathematics from Brown University.

PETER HUGHES

Peter Hughes is the NASA Goddard Space Flight Center's (GSFC) acting chief technologist, and (acting) head of the Goddard Technology Management Office (GTMO)/Code 502 and also serves as chair of the Technology Federation. The Technology Federation serves as a pro-active advisory board to Goddard's management about centerwide technology issues and serves to stimulate innovation and bridge the science and engineering communities at GSFC. The Technology Federation brings together the many faces of GSFC and is comprised of representatives from GTMO, the Goddard Sciences and Exploration Directorate, Flight Programs and Projects Directorate, Wallops Flight Facility, and the AETD Engineering Divisions through their respective assistant chiefs for technology (ACTs).

Previously, Mr. Hughes was the assistant chief for technology in the Information Systems Division (ISD)/Code 580 at GSFC. In this position, he led the ISD's Strategic Technology Planning, served as the ISD liaison and point of contact for advanced technology, and managed the ISD advanced technology program in preparation for GSFC's next generation science missions.

Mr. Hughes has also served as a technology systems engineer in the Mission Implementation and Technology Management Office/Code 510.1. In this position, he initiated and led the Flight Testbed for Innovative Mission Operations and two satellite control center automation initiatives—the EUVE Automated Payload Operations Control Center (APOCC) and GRO Reduced Operations by Optimizing Techniques and Technologies (ROBOTT) projects. Mr. Hughes also served as the Mission Technologist for the Extreme UltraViolet Explorer (EUVE) and Hubble Space Telescope Ground System (for the Vision 2000 Project).

In addition, he served as team lead for one of the Mission Operations and Data Systems Directorate's re-engineering team.

Mr. Hughes previously worked in the Data Systems Technology Division, where he served as the technical lead and Project Manager of the GenSAA System and for which he holds a U.S. Patent. He also designed and implemented the CLEAR System, the first real-time expert system to monitor a low-earth-orbit satellite. Additionally, he supported a number of other initiatives investigating advanced technologies in Artificial Intelligence, Software Engineering, and Human Factors research.

Mr. Hughes received his B.S. in computer science from the College of William and Mary and an M.S. in computer science from Johns Hopkins University. In May 2004, he received an M.S. in management of technology from the University of Pennsylvania's Executive Masters for Technology Management (EMTM) program, a joint program sponsored by the Wharton Business School and the SEAS School of Engineering. He was sponsored by the NASA Fellowship Program.

TREVOR O. JONES

Trevor O. Jones is the chairman and founder of BIOMEC Inc., an entrepreneurial company founded in 1998 engaged in the development and commercialization of biomedical engineered devices and products.

After seven years, Mr. Jones retired from the board of directors of Echlin, Inc. in June 1998, where he served in a number of capacities as chairman, vice chairman, chief executive officer, and chairman of its European Advisory Committee.

Mr. Jones was appointed chairman of the board of Libbey-Owens-Ford Co. in 1987, and assumed the additional positions of president and chief executive officer in May 1993. Mr. Jones retired from Libbey-Owens-Ford in 1994 but remained a member of the board of directors including chairman of their Salary and Bonus Committee until 1997.

From 1978 to 1987, Mr. Jones was an officer of TRW, Inc. He joined TRW in 1978 as vice president, Engineering, Automotive Worldwide Sector and in 1979 he formed TRW's Transportation Electronics Group and was appointed its group vice president and general manager. His responsibilities included activities in the United States, United Kingdom, and Japan. In 1985, his responsibilities were further expanded to include Sales, Marketing, Strategic Planning, and Business Development activities for the entire Automotive Sector.

From 1959 to 1978, Mr. Jones spent 19 years with General Motors. His last position there was director of General Motors Proving Grounds, a post to which he was appointed in 1974. From 1959 to 1970, Mr Jones was involved in General Motors' aerospace activities at the Delco Electronics Division. During this period, he directed many major programs, including the B-52 bombing navigational system production program, advanced military avionic systems, and the Apollo lunar and command module computers. In 1969, he was selected to direct the application of aerospace technology to automotive safety and electronics systems. He became the director, Automotive Electronic Control Systems, a newly organized group at General Motors Technical Center in 1970 and was appointed director, Advanced Project Engineering in 1972. In this capacity, he directed many major vehicle, engine, and component development programs.

In 1982 he was elected a member of the National Academy of Engineering and was cited for "leadership in the application of electronics to the automobile to enhance its mechanical performance." He has been a member of a number of National Research Council (NRC) study committees, including "National Interests in an Age of Global Technology," "Safety Research for a Changing Highway Environment," "Engineering as an International Enterprise," "Competitiveness of the U.S. Automotive Industry," and "Time Horizons and Technology Investments." In 1993, Mr. Jones chaired the National Academy of Engineering Committee on the effects of products liability law on innovation.

From 1994 to 2000, Mr. Jones chaired the National Research Council's Standing Committee for the Partnership for a New Generation Vehicle, which is often referred to as the "80 mile per gallon super car". Mr. Jones continues to be active in fuel cell developments and is a member of UTC's Fuel Cell Advisory Committee and a member of the Executive Committee of the Ohio Fuel Cell Coalition.

NICK KARANGELEN

Nick Karangelen is the president and founder of Trident Systems Incorporated, which provides technology solutions to industry and government clients in a broad spectrum of application areas and conducts research initiatives in advanced systems engineering methods and tools. As president, he directs strategic investment in emerging technologies, oversees Trident's ongoing research initiatives, and leads the continuing expansion and refinement of Trident's world-class complex systems engineering capabilities. He is a 1976 graduate of the U.S. Naval Academy and served as a naval officer in the U.S. nuclear submarine force during the Cold War. Trident celebrated its 20th year in business in 2005.

MAX V. KIDALOV

Max V. Kidalov serves as Counsel to the U.S. Senate Committee on Small Business and Entrepreneurship, chaired by Senator Olympia J. Snowe (R-ME). At the Committee, he is responsible for legislative and oversight matters concerning all federal procurement and technology programs affecting small business, as well as waste, fraud, and abuse issues.

Mr. Kidalov has an extensive background in the fields of government contracts and federal claims. He represented clients in matters concerning bid protests, contract disputes, and procurement integrity at the Washington, D.C., law firm of Spriggs & Hollingsworth and consulted on procurement strategies with former Congressman Mark Siljander (R-MI).

Prior to entering private practice, Mr. Kidalov was a two-term law clerk to Chief (now Senior) Judge Loren A. Smith of the U.S. Court of Federal Claims, where he handled numerous procurement and regulatory contracts cases as well as other monetary federal claims. In public service, Mr. Kidalov also worked in the office of Governor David Beasley (R-SC) and at the U.S. Senate Committee on the Judiciary on the staff of the late Senator Strom Thurmond (R-SC).

Mr. Kidalov is a former vice chairman of the Bid Protest Committee for the American Bar Association Section of Public Contract Law, and has written numerous legal and policy articles concerning government contracts, takings of private property rights, and other federal claims. He is a member of the bars of the Supreme Court of South Carolina, the U.S. Court of Federal Claims, and the U.S. Court of Appeals for the Federal Circuit.

Mr. Kidalov received his B.S. *cum laude* and his J.D. degrees from the University of South Carolina, and is presently an L.L.M. degree candidate in the Government Procurement Program at the George Washington University.

STEPHEN LEE

Stephen Lee is currently the director of organic chemistry at the U.S. Army Research Office and an adjunct faculty member in chemistry at the University of North Carolina-Chapel Hill. The Army Research Office program includes basic research directed towards hazardous materials management including basic research in decontamination. The research is focused on technologies needs of the warfighter for sensing, decon, and protection. He received a B.S. degree from Millsaps College in chemistry and biology and a Ph.D. from Emory University in physical organic chemistry. Dr. Lee was also a Chateaubriand Fellow at the Université Louis Pasteur in Strasbourg, France, studying origin of life chemistry.

PETER LEVINE

Peter Levine has served as minority counsel to the Senate Armed Services Committee since January 2003 and from 1996 to 2001. In 2001 and 2002, Mr. Levine served as general counsel of the Committee. In both positions, Mr. Levine has been responsible for providing legal advice on legislation, nominations, and other matters coming before the Committee. He also advises members of the Committee on acquisition policy, environmental policy, and defense management issues impacting the Department of Defense.

Previously, Mr. Levine served as counsel to Senator Carl Levin of Michigan (1995-1996), and counsel to the Subcommittee on Oversight of Government Management of the Senate Governmental Affairs Committee (1987-1994). In his capacity as counsel to Senator Levin and to the Oversight Subcommittee, Mr. Levine was responsible for efforts to overhaul the lobbying disclosure laws and streamline the federal procurement system. Mr. Levine was a key participant in a broad array of legislative measures, including the Information Technology Management Reform Act of 1996, the Lobbying Disclosure Act of 1995, the Senate gift reform resolution, the Federal Acquisition Streamlining Act of 1994, the Clean Air Act of 1990 (mobile sources provisions), the Whistleblower Protection Act of 1989, the Ethics Reform Act of 1989, and the Office of Federal Procurement Policy Reauthorization Act of 1988. Mr. Levine has also handled a number of oversight matters, including the 1987 congressional investigation of the Wedtech Corporation, congressional efforts to encourage broader use of commercial items and commercial practices in government procurement, and efforts to identify and eliminate wasteful practices in the management of defense inventory.

Prior to joining the Senate staff, Mr. Levine was an associate at the law firm of Crowell & Moring. Mr. Levine graduated *summa cum laude* from Harvard College in 1979 and *magna cum laude* from Harvard Law School, where he was an editor of the *Harvard Law Review*, in 1983.

MICHAEL MCGRATH

Michael McGrath was appointed as the Deputy Assistant Secretary of the Navy for Research, Development, Test and Evaluation in February 2003. His role is to aggressively drive new technologies from all sources across Navy and Marine Corps platforms and systems, and to develop programs to bridge the gap in transitioning from science and technology to acquisition. He is also responsible for developing new ways to integrate Test and Evaluation (T&E) with the evolutionary acquisition process.

Prior to his appointment to this position, Dr. McGrath spent 5 years as vice president for government business at the Sarnoff Corporation, a leading R&D company with both commercial and government clients. He was responsible for program development across all Sarnoff business units to meet government needs for innovative dual-use technologies in sensors and microelectronics, networking and information technology, and bio-technology.

Dr. McGrath has 28 years of prior government experience. His early career was in weapon system logistics planning and management, first at the Naval Air Systems Command, and later in the Office of the Secretary of Defense, where he developed policies for Integrated Logistics Support and reviewed implementation in major weapon system acquisition programs in all three military departments.

Dr. McGrath was appointed to the Senior Executive Service in 1986 as director of the OSD CALS Office, where he guided the Computer-aided Acquisition and Logistics Support program from its inception. Five years later he became the assistant director for manufacturing in DARPA's Defense Sciences Office, where he managed programs in Agile Manufacturing, Electronic Commerce Resource Centers, and Affordable Multi Missile Manufacturing. He also served in leadership positions for several DoD-wide initiatives to improve manufacturing and reduce the cost of defense systems. In 1996-1997 he served as the Assistant Deputy Under Secretary of Defense (Dual Use and Commercial Programs), where he directed the Commercial Technology Insertion Program, the Commercial Operating and Support Savings Initiative, and the Department's Title III industrial base investments.

Dr. McGrath holds a B.S. in space science and applied physics (1970) and an M.S. in aerospace engineering (1972) from Catholic University, and a doctorate in operations research from George Washington University (1985). He has been active in several industry associations and study groups, including studies by the Defense Science Board and the National Research Council.

RICHARD MCNAMARA

Richard McNamara has provided technical direction and leadership for complex acoustic, mechanical, and electronic combat and weapon systems for more than 30 years, holding positions in the Naval Underwater Warfare Center (NUWC), Naval Sea Systems Command (NAVSEA), and Assistant Secretary of the Navy (Research, Development & Acquisition) (ASN(RDA)), and Program Executive Officer for Submarines (PEOSUB).

Mr. McNamara's civilian service began in a work-study program between the Naval Underwater Systems Center (NUSC) and Northeastern University. After graduation from Northeastern in 1972 with a B.S. degree in mechanical engineering he returned to NUSC to plan and direct numerous sonar and towed array test programs.

In 1977 Mr. McNamara joined NAVSEA, and in 1983, as the Head of the PMS409 Combat Control Systems Engineering Branch, Mr. McNamara contributed to the successful submarine fleet introduction of TOMAHAWK and Over the Horizon Targeting (OTH-T) and initial testing of ADCAP torpedoes.

After serving 10 months as a legislative fellow for Senator Gramm from Texas in 1986, Mr. McNamara became deputy program manager for Submarine Combat Systems (PMS409). From 1991-1992 Mr. McNamara served as the technical director for the PEO for Submarine Combat and Weapon Systems (SCWS), acting as the senior technical advisor for Flag level decisions regarding submarine programs. In 1993 Mr. McNamara assumed responsibilities for all combat system activities for the Virginia Class SSN, spearheading the acquisition of the Virginia Class SSN Command, Control, Communications, and Intelligence (C^{3}I) System. In 1997 Mr. McNamara was assigned as the deputy program manager for Submarine Electronic Systems Program Office (PMS401), providing leadership and management direction for the activities of the NSSN C^{3}I System and ancillary submarine electronics systems

Mr. McNamara was selected as the deputy program manager for the ACAT 1D Virginia Class Submarine Program Office (PMS450) in December 1998. He now is responsible for managing the acquisition programs for PEO Submarines as executive director.

Mr. McNamara has received a variety of awards throughout his career including Secretary of the Navy Competition Advocate Award, Association of Scientists and Engineers Silver Medal, and David Packard Award of Acquisition Excellence, among others.

ANTHONY C. MULLIGAN

Anthony C. Mulligan is the president and chief executive officer of Advanced Ceramics Research. He was one of the original founders of the company in 1989. ACR is one of a handful of companies to have achieved a perfect 100 commercial activity index (CAI) in the Department of Defense SBIR program. In 2000 and 2004, ACR was featured by the U.S. Navy as a SBIR role model company. ACR has also been featured as a role model success story by the U.S. Department of Energy (2001 and 2002) and by the National Aeronautics and Space Administration (1997 and 2004). ACR has received three R&D 100 awards, The Tom Brown Entrepreneur Award, the Arizona Innovation Award, and was also a team member of the "ACTD Team of the Year Award" for the Expendable UAV ACTD project (2003). ACR has also created a Manufacturing Joint Venture Company, Advanced Ceramics Manufacturing, LLC, on the Tohono O'Odham Reservation in Southwest Arizona. Since 1989 ACR has generated 60 granted U.S. Patents, with Mr. Mulligan receiving 25.

Mr. Mulligan was also the founder and a principal of Revdyne, Inc., a company that manufactured products for developmentally disabled individuals. Mr. Mulligan was also the founder and principal of a successful pet products manufacturing company, which manufactured and sold products in high volumes to major U.S. department stores including Kmart, Walgreens, Albertsons, Frys, PetSmart, and Ames.

Mr. Mulligan is on the Industrial Advisory Counsel to the Aerospace and Mechanical Engineering Department at the University of Arizona. For two years he served on the board to the Arizona Manufacturing Extension Partnership (MEP). He has served as chairman to the Small Manufacturing Executives of Tucson for three years. He is a member of ASM, the American Ceramics Society, the Society of Manufacturing Engineers (SME), and is on the Structural Materials Committee for TMS where he served a term as chairman to the Young Leaders Committee.

Mr. Mulligan received a B.S. in mechanical engineering from the University of Arizona in 1988.

KENT MURPHY

Kent Murphy is chief executive officer and founder of Luna Innovations, Inc., a Blacksburg, VA-based, employee-owned corporation. He also serves on the Committee for the National Academies SBIR study.

Luna is a business development company that identifies significant market opportunities, builds promising intellectual property portfolios and prototypes, and delivers them into highly differentiated commercial applications. Since 2000, Luna has spun-off five new companies focusing on the areas of manufacturing process control, nanomaterials, proteomics and analytical instrumentation, petroleum monitoring systems, and integrated wireless sensing systems. Luna is a two-time Tibbetts award winner.

Dr. Murphy is formerly a tenured professor in Virginia Tech's Bradley Department of Engineering. He has over 35 patents which have generated hundreds of millions of dollars in product revenue. In 2001, he was named Virginia SBIR

Entrepreneur of the Year, and this year was recognized by the Governor and Science Museum of Virginia as Virginia's Outstanding Industrialist of the Year. Dr. Murphy is a founding member of the Virginia Research and Technology Advisory Commission, appointed by the Governor, and continues to serve today. He is also a member of the Greater Washington Board of Trade Virtual Incubator Action Committee, and the Potomac Tech Task Force. In May of 2003, Dr. Murphy gave testimony at the Full Committee Hearing of the Senate Committee on Commerce, Science, and Transportation on S.189, the 21st Century Nanotechnology Research and Development Act, a $3.4 billion nanotechnology funding bill.

ROBERT M. PAP

Robert M. Pap is the president and cofounder of Accurate Automation Corporation (AAC). AAC, founded in 1985, is a woman-owned, high-tech developer of unmanned aircraft and guided missiles. Accurate is known for its expertise in intelligent control systems and signal processing technology. AAC is a world leader in plasma aerodynamic research. Its neural network hardware is being used around the world. The Accurate facilities support UAV development and manufacturing. AAC has world-class facilities for plasma research. Its ground stations support flight tests for its jet-powered aircraft including LoFLYTE and the X-43A-LS.

Currently, Accurate Automation Corporation has the following commercial product areas: Unmanned Aerial Vehicles; Jet engines (150-200 lb thrust range); Flight control and telemetry systems; Neural Network Processor (NNP®); and Aircraft Safety and Security (Imagery/Telemetry) Systems.

JOHN A. PARMENTOLA

John A. Parmentola is the director for Research and Laboratory Management for the U.S. Army. In this position, he is responsible for the Army Basic Research Program and the Applied Research programs of the Army Research Laboratory, Army Research Institute, Corps of Engineers, and Simulation, Training and Instrumentation Command. In addition, his responsibilities encompass Environmental Quality Technology, Manufacturing Technology, Small Business Innovative Research, Dual Use Science and Technology, and Army High Performance Computing programs with an annual budget of approximately $750 million. Dr. Parmentola also oversees laboratory management policy for all Army laboratories and research, development, and engineering centers.

Before coming to the Army, Dr. Parmentola was the science and technology advisor to the chief financial officer of the U.S. Department of Energy, where he was responsible for providing technical, budgetary, and programmatic advice to senior management for over $7 billion in science and technology investments.

This responsibility included Defense, Non-proliferation, Science, Fossil Energy, Energy Efficiency, Nuclear Energy, and Environmental programs. Prior to joining the U.S. Department of Energy, he was the co-founder of the Advanced Systems and Concepts Office of the newly formed Defense Threat Reduction Agency.

Dr. Parmentola has been a principal scientist at the MITRE Corporation, where he has worked in the area of arms control verification technology, strategic offense-defense integration, and strategic command, control, and communications associated with the Cheyenne Mountain Upgrade Program. Earlier in his career, he was executive director of the Panel on Public Affairs of the American Physical Society, a postdoctoral fellow with the Program of Science and Technology for International Security at MIT, and a postdoctoral fellow with the Laboratory for Nuclear Science of MIT. In the field of science, technology, and public policy, he has been a fellow of the Roosevelt Center for American Policy Studies and an Alfred P. Sloan Research Fellow at the Center for Science and International Affairs of the John F. Kennedy School of Government at Harvard University.

Dr. Parmentola has published more than 50 scientific papers and articles in science and technology policy, and an authoritative book on space defense. He has been the recipient of the Alfred Raymond Prize, the Sigma Xi Research Award, has been an Andrew Mellon Postdoctoral Fellow, and an Alfred P. Sloan Fellow. Recently, he was the Air Intelligence Agency nominee for the R.V. Jones Central Intelligence Agency Award and has been awarded the Outstanding Civilian Service Award for his exemplary dedication to public service and his numerous contributions to the U.S. Army.

Dr. Parmentola was born in the Bronx, New York, and received his B.S. degree in physics from Polytechnic Institute of Brooklyn in 1971 and his Ph.D. in physics from the Massachusetts Institute of Technology, Cambridge, Massachusetts in 1977.

MARIO RAMIREZ

Mario Ramirez is the F-35/Joint Strike Fighter Small Business Officer for Lockheed Martin Aeronautics.

CARL G. RAY

Carl G. Ray is the NASA program executive for the NASA Small Business Innovative Research and Small Business Technology Transfer (SBIR/STTR) Element Programs. He is the source selection official (SSO) for these programs and as such responsible for all final awards under these programs, and their agency-level policy and strategic oversight. These element programs are part of the Innovative Partnerships Program under the Exploration Systems Mission Directorate.

Mr. Ray is also the NASA technical director for NASA's renowned "NASA Tech Briefs" magazine.

MARK REDDING

Mark Redding is the president of Impact Technologies. He has been participating in the SBIR program since 1990. While vice president at a 25-person mechanical engineering firm during the 1990's, the company was awarded three DoD Phase I contracts and two Phase II contracts. Since he co-founded Impact Technologies in 1999, Impact has been awarded over 45 Phase I and 30 Phase II DoD SBIRs, making it the top award winner in New York State each of the past three years. The awards have included most DoD agencies (Navy, ONR, Army, Air Force, DARPA, and AFOSR) in addition to a few NASA and Department of Transportation SBIRs. The resulting technologies have been transitioned to more than a dozen military platforms, including the JSF F135/F136.

EARLE RUDOLPH

Earle Rudolph is vice president for business and strategy development in the ATK Mission Research Group. He spent 23 years in the Navy with tours as the first deputy program manager for JDAM under the Air Force program manager. He also served as professor of Information Based Warfare at the National Defense University.

After leaving the Navy, Mr. Rudolph joined Texas Instruments, "New Business Strategy Group", chartered to develop advanced systems value propositions for the Defense Systems Group. When DSEG was sold to Raytheon he moved to the Raytheon Missile Systems Advanced Programs Division as a program/capture manager. In 2000 he moved to Draper Laboratories Washington office as the business development director. Mr. Rudolph joined ATK as vice president for business and strategy development in 2002 and currently serves in that capacity in the recently acquired ATK Mission Research Group.

MARK D. STEPHEN

Col. Mark D. Stephen is the chief, Science and Technology Division, Deputy Assistant Secretary (Science, Technology & Engineering), Pentagon, Washington D.C.—the Air Force Secretariat focal point for policy, programming, planning, budgeting, and Congressional matters concerning the Air Force science and technology program.

In his previous assignment, Col. Stephen was serving as the acting director of the Directed Energy Directorate, Air Force Research Laboratory, Kirtland Air Force Base, NM, the organization responsible for advancing all Air Force high-energy laser, high-power microwave, and other directed energy technologies. He was commissioned as a distinguished graduate through the Reserve Officer Training Corps in 1978. His first assignment was to the Air Force Institute of Technology where he earned a master's of science degree in engineering physics. He is a level three acquisition manager and has earned the Air Force Master Space Badge, Master Acquisition Badge, Senior Missile Badge, and the Information Management Badge. He is also a fully trained joint specialty officer.

Col. Stephen earned a B.S. in physics, University of South Carolina, Columbia, South Carolina, 1978; an M.S. in engineering physics, Air Force Institute of Technology, Wright-Patterson Air Force Base, Ohio, 1979; and completed Squadron Officers School (outstanding contributor), Maxwell Air Force Base, Alabama, 1983; Program Management Course, Defense Systems Management College, 1989; Air Command and Staff College (distinguished graduate), Maxwell Air Force Base, Alabama, 1990; Armed Forces Staff College, Norfolk, Virginia, 1990; Air War College, seminar, Kirtland Air Force Base, New Mexico; 1995; Executive Refresher Course, Defense Acquisition University, 2001.

Col. Stephen's major awards and decorations include the Defense Meritorious Service Medal, Meritorious Service Medal (four oak leaf clusters), Joint Service Commendation Medal, Air Force Commendation Medal, Joint Service Achievement Medal, Air Force Achievement Medal, and Military Outstanding Volunteer Service Medal.

JAMES TURNER

James Turner currently serves as the chief Democratic counsel for the House Committee on Science where he works across the broad range of issues concerning science, technology, energy, and space exploration that characterize the Science Committee's legislative agenda. Having served on the professional staff of the Committee for approximately 20 years, Mr. Turner brings an exceptional perspective on the legislative history and process in the area of U.S. science and technology policy. He is widely recognized in U.S. policy circles for his experience, effectiveness, and willingness to find common ground on complex issues of national policy.

For the 10 years prior to the Republican's winning the majority in the Congress, Mr. Turner served as the Committee's senior staff member for technology policy including four years as staff director for the Subcommittee on Technology, as well as Subcommittee Legal Counsel. During the late 1970s and early 1980s, Mr. Turner worked on the Committee's Republican staff as minority energy counsel.

During his years on the Committee, Mr. Turner has played a major role in the drafting and negotiation of numerous legislative initiatives, Congressional reports, and hearings on a wide variety of topics. These include the international competitiveness of U.S. industry, environmental and energy research and development, trade and technology policy, intellectual property, standards, and technology transfer.

Mr. Turner's experience also includes work outside the Science Committee. He spent three years working for Wheelabrator-Frye, two years for Congressman Gary Myers, two years for the State of Connecticut, and shorter periods in the executive branch with NASA and the FAA. He holds degrees from Westminster College and from the Universities of Georgetown and Yale. Reflecting his accomplishments, he was selected to attend Harvard University's Senior Managers in Government Program.

TRACY VAN ZUIDEN

Major Tracy Van Zuiden is the Technology Transition lead for the Joint Strike Fighter Office in Arlington, Virginia. This program will develop and produce the next generation strike warfare weapon system for the United States Navy, Marines, Air Force; for the Royal Navy and Air Force; and for Italy, the Netherlands, Turkey, Canada, Australia, Denmark, Norway, as well as other foreign military sales customers.

In his current position Major Van Zuiden is responsible for roadmapping technologies to meet current and future F-35 capability requirements.

Major Van Zuiden was commissioned in 1989 through the ROTC graduating from the University of Central Florida. During his career he has held numerous positions in both acquisitions and aircraft maintenance to include F-15 flight-line and intermediate maintenance, C-5 modernization programs, in-country logistics liaison to the Royal Saudi Air Force, and in the Air Force Program Executive Office for Airlift & Trainer.

JOHN P. WASZCZAK

John P. Waszczak is the director of Advanced Technology and SBIR/ STTR, for Raytheon Missile Systems (RMS). RMS designs, manufactures, and services tactical weapon systems for the United States and allied governments.

Most recently, Dr. Waszczak served as the RMS Product Line deputy for Guided Projectiles, which included ERGM (Extended Range Guided Munition) for the Navy, and Excalibur (XM982) for the Army and Marine Corps.

Dr. Waszczak's career with Raytheon (formerly General Motors Hughes and General Dynamics) spans over 30 years. He served as the director of Materiel Operations for GM Hughes/Raytheon Missile Systems, responsible for the annual procurement of $1B in materials and services.

At GM Hughes, Dr. Waszczak served as the deputy managing director of Hughes UK Ltd., while stationed in London, England. He previously served as the deputy director of Tomahawk Cruise Missile Programs. Dr. Waszczak was the recipient of the Hughes Electronics Malcolm R. Currie Innovation Award in 1996.

At GD Convair, Dr. Waszczak held numerous positions, including director of Tomahawk Cruise Missile Production Programs, director of division planning, director of strategic planning, director of Zero Defect Management Administration, and director of facility management.

Dr. Waszczak led the GD Convair Division transition team during the split of the Convair Division into the Convair & Space Systems Divisions. At GD he helped develop a new product line, Energy Systems, serving sequentially as marketing manager, chief engineer, program manager, and then manager of Energy Systems. John also served as principal investigator, project manager, and program manager in advanced composite materials research.

Dr. Waszczak was special assistant to the U.S. Department of Transportation's Assistant Secretary for Policy and International Affairs while a member of the President's Executive Exchange Program during the Carter Administration.

Dr. Waszczak is an alumnus of Carnegie-Mellon University where he obtained his B.S., M.S., and Ph.D. degrees in mechanical engineering with a minor in economics. His thesis work focused on automated design procedures for advanced composite materials, sponsored by the U.S. Air Force and General Dynamics.

Dr. Waszczak was commissioned in the U.S. Army Corps of Engineers and was honorably discharged as Captain. He is a member of Pi Tau Sigma, Tau Beta Pi and Phi Kappa Phi honorary fraternities, and the National Management Association.

CHARLES W. WESSNER

Charles W. Wessner is the director of the National Academies study of SBIR. He is a policy advisor recognized nationally and internationally for his expertise on innovation policy, including public-private partnerships, entrepreneurship, early-stage financing for new firms, and the special needs and benefits of high-technology industry. He testifies to the U.S. Congress and major national commissions, advises agencies of the U.S. government and international organizations, and lectures at major universities in the United States and abroad. Reflecting the strong global interest in innovation, he is frequently asked to address issues of shared policy interest with foreign governments, universities, and research institutes, often briefing government ministers and senior officials. He has served as an advisor to the 30-nation OECD Committee on Science and Technology Policy, the Mexican National Council on Science and Technology, and the National Technology agencies of Finland (TEKES) and Sweden (VINNOVA), and is a member of the Norwegian Technology Forum. He also serves as a member of the Prime Minister of Taiwan's Science and Technology Advisory Group and is a member of the Lithuanian Prime Minister's International Innovation Advisory Committee and the board of the Vilnius Sunrise Valley S&T Park. Most recently, he was named to the U.S.-Russian Council on Innovation, established by Presidents Putin and Bush.

Dr. Wessner's work focuses on the linkages between science-based economic growth, entrepreneurship, new technology development, university-industry clusters, regional development, small firm finance, and public-private partnerships. His program at the National Academies also addresses policy issues associated with international technology cooperation, investment, and trade in high-technology industries. Currently, he directs a series of studies centered on government measures to encourage entrepreneurship and support the development of new technologies. Foremost among these is a Congressionally mandated study of the $2 billion Small Business Innovation Research program. A major new comparative study of National Innovation Policies is now underway.

Dr. Wessner is an Ameritech Research Fellow at the Indiana University School of Public and Environmental Affairs where he also serves on the Visiting Committee. He teaches as an adjunct professor at George Washington University's Elliott School of International Affairs, is a research professor at the Max Planck Institute for Research into Economic Systems in Jena, Germany, and is a visiting professor at the University of Vilnius, Lithuania. As a recognized expert on U.S. innovation policy, he has spoken before the House Science, Small Business, and Armed Services Committees and the Senate Small Business Committee, and to the President's Council of Advisors on Science and Technology (PCAST). Reflecting the growing overseas interest in understanding U.S. innovation policy, he has addressed parliamentarians from the 33 EUREKA countries in the Danish Parliament, the Canadian Prime Minister's Advisory Council on Science and Technology, the Board of the European Investment Bank, CDU members of the Education and Research Committee in the Bundestag, and members of the Swedish and Lithuanian Parliaments.

KEVIN WHEELER

Ms. Kevin Wheeler is the deputy Democratic staff director for Senator John F. Kerry (D-Mass.) on the Committee on Small Business & Entrepreneurship. She handles legislation and policy that covers the Small Business Administration's credit and venture capital programs, hi-technology programs for small businesses, and SBA's budget and appropriations. Before joining Senator Kerry's staff in 1998, she spent three years as the assistant editor of *Business New Haven*, a regional business journal in Connecticut. Prior to that, she worked for Senators Lloyd Bentsen and Bob Krueger of Texas, and Bill Curry, a Democrat who ran for Governor of Connecticut in 1993 and later became counsel in the Clinton White House.

JOHN WILLIAMS

John Williams is the program manager for the Navy's Small Business Innovation Research Program and the Small Business Technology Transfer Program. He has been with the Navy for 17 years, spending time at the Naval Surface Warfare Center, Naval Sea Systems Command, and his last nine years with the Office of Naval Research. During his career with the Navy he has worked with the Navy's Manufacturing Technology Program, the Navy and Private Shipyards, and the National Shipbuilding Research Program, and he has been the program officer for multiple SBIR projects.

In 1996 Mr. Williams joined ONR and became the deputy to the Navy SBIR program. His main focus has been to transition the Navy SBIR program into a paperless environment, to manage the STTR program, and most recently to increase the commercialization or more accurately the transition of Navy SBIR and STTR technologies into the fleet. In 2000 John initiated the Transition Assistance Program (TAP), a 10-month program designed to educate and assist all Navy Phase II awardees in the technology transition process. Recently this effort was expanded to include the Primes Initiative, which is focused on increasing the involvement of DoD prime contractors into the SBIR program. TAP concludes with the Navy Opportunity Forum, scheduled for the first Monday and Tuesday in May, and this year's event had over 700 representatives from DoD primes, program offices, and small business.

The Navy has led the way at increasing the involvement of acquisition program offices and major defense contractors in the SBIR program with the goal that developing closer partnership between our nation's small high-tech firms and these organizations will ultimately increase the transition of SBIR- and STTR-funded technologies. This has proven true, and the Navy has the highest record of Government Phase III contracts across the DoD. Mr. Williams has a degree in mechanical engineering from the University of Maryland, College Park and a master's in engineering management, marketing of technology from the George Washington University.

Appendix B:
Participants List
14 June 2005 Symposium

Zoltan Acs
University of Baltimore

Alan Anderson
National Research Council

Clare Asmail
National Institute of Standards and
    Technology

Allen Baker
Vital Strategies

Tabitha Benney
National Research Council

Robert Berger
Department of Energy, (ret)

Richard Bissell
National Research Council

Jeff Bond
Association for Manufacturing
    Technology

Tony Bower
RAND

Edsel Brown
Small Business Administration

Michael Caccuitto
Department of Defense

Peter Cahill
BRTRC

Jeffrey Carroll
Innovative Defense Strategies

Julianne Carroll
Senate Committee on Small Business

Richard Carroll
Innovative Defense Strategies

Speakers are in italics.

171

Tom Cassin
Materials Science Corporation

Victor Ciardello
Department of Defense

McAlister Clabaugh
National Research Council

Major Clark
Small Business Administration

Charles Cleland
Department of Agriculture

Ronald Cooper
Small Business Administration

Candice Cotton
Office of Senator Enzi

Thomas Crabb
Orbitec

Diane DeVaul
Northeast-Midwest Institute

David Dierksheide
National Research Council

Aaron Druck
Washington CORE

Rosalie Dunn
National Institutes of Health

Kay Etzler
National Institutes of Health

David Finifter
College of William and Mary

Kevin Finneran
The National Academies

Michael Fitzgerald
Technology Tree Group

James Gallup
Environmental Protection Agency

Jacques S. Gansler
University of Maryland

Robin Gaster
North Atlantic Research

Mike Gerich
Office of Management and Budget

Jere Glover
Small Business Technology Coalition

Vinod Goel
The World Bank

Carlos Gorrichategui
Technology Tree Group

Margaret Grabb
National Institutes of Health

Bill Greenwalt
Senate Committee on Armed Services

Steve Guilfoos
U.S. Air Force

Paul Hauler
Dawnbreaker

Chris Hayter
NACFAM

Anne Heath
National Institutes of Health

Richard H. Hendel
Boeing Corporation

Claude Hennessey
U.S. Air Force

Charles J. Holland
Department of Defense

Peter Hughes
NASA Goddard Space Flight Center

Olwen Huxley
House Committee on Science

Hiroshi Ikukawa
Embassy of Japan

Heidi Jacobus
Cybernet Systems

Larry James
Department of Energy

George Johnson
National Institutes of Health

Trevor Jones
BIOMEC, Inc.

James Kadtke
Office of Senator Warner

Nick Karangelen
Trident Systems

Nicholas Karvonides
Manufacturing Extension Partnership

Max V. Kidalov
Senate Committee on Small Business
and Entrepreneurship

Eli Kintisch
Science Magazine

Kristopher Koenen
Department of Commerce

Andrew Kovacs
House Committee on Science

Robert Lautrup
House Committee on Armed Services

Emmanuel LePerru
Embassy of France

Stephen Lee
U.S. Army Research Office

Peter Levine
Senate Committee on Armed Services

Lynn Luethke
National Institutes of Health

Peter Magumdar
Department of Defense

Joshua Mandell
Embassy of Great Britain

Goran Marklund
Embassy of Sweden

Elaine McCusker
Senate Committee on Armed Services

Neil McDonald
Federal Technology Watch

Michael McGrath
U.S. Navy

Richard McNamara
U.S. Navy

Edward Metz
Department of Education

Stephen Mincemoyer
Innovative Defense Strategies

Duncan Moore
University of Rochester

Francisco Moris
National Science Foundation

Anthony C. Mulligan
Advanced Ceramics Research

Kent Murphy
Luna Innovations

Kesh Narayanan
National Science Foundation

Carl Nelson
Carl Nelson Consulting

Susan Nichols
U.S. Army

Markku Oikaraienen
Embassy of Finland

Harold Olsen
Anteon

Diane Palmintera
Innovation Associates

Robert M. Pap
Accurate Automation Corporation

John A. Parmentola
U.S. Army

Karen Pera
SOCOM

Susan Pucie
National Institutes of Health

Mike Quear
House Committee on Science

Lisette Ramcharan
Embassy of Canada

Mario Ramirez
Lockheed Martin

Frank Ramos
Department of Defense

Carl G. Ray
National Aeronautics and Space
Administration

Mark Redding
Impact Technologies, LLC

Jean Reed
House Committee on Armed Services

Marshall Reffett
Department of Commerce

Volker Rieke
Embassy of Germany

Frank Rucky
Missile Defense Agency

James Rudd
National Science Foundation

Earle Rudolph
ATK

Craig Rutler
U.S. Army

Arun Seraphin
Senate Committee on Armed Services

Jenny Servo
Dawnbreaker

Devanand Shenoy
Naval Research Lab

Hideo Shindo
NEDO

Kathleen Shino
National Institutes of Health

Stephanie Shipp
National Institute of Standards and
    Technology

Sujai Shivakumar
National Research Council

Victor Shulepov
Embassy of Ukraine

Robert Sienkiewicz
National Institute of Standards and
    Technology

Dennis Sorensen
Office of Naval Research

Barbara Staals
Embassy of the Netherlands

Mark D. Stephen
U.S. Air Force

Nigel Stephens
Senate Committee on Small Business
    and Entrepreneurship

Katie Stevens
Office of Congressman Baird

Stephen Sullivan
U.S. Navy

Maurice Swinton
Small Business Administration

Roland Tibbetts
National Science Foundation, (ret.)

Andrew Toole
Rutgers University

Lynn Torres
Office of Naval Research

Etienne Toussaint
House Committee on Science

James Turner
House Committee on Science

Carol Van Wyk
U.S. Navy

Tracy Van Zuiden
U.S. Air Force

Christine Villa
BRTRC

Starnes Walker
Office of Naval Research

Wilson Wang
Office of Senator Lieberman

Tab Wilkins
Washington Technology Center

John P. Waszczak
Raytheon Company

John Williams
U.S. Navy

Eric Webster
House Committee on Science

Michelle Willis
U.S. Navy

Charles W. Wessner
National Research Council

Jim Woo
Interscience

Kevin Wheeler
Senate Committee on Small Business
    and Entrepreneurshp

Grazyna Zebrowska
Embassy of Poland

Appendix C:

Bibliography

Acs, Zoltan and David B. Audretsch. 1988. "Innovation in Large and Small Firms: An Empirical Analysis." *The American Economic Review* 78(4):678-690. September

Acs, Zoltan and David B. Audretsch. 1991. *Innovation and Small Firms.* Cambridge, MA: The MIT Press.

Alic, John A., Lewis Branscomb, Harvey Brooks, Ashton B. Carter, and Gerald L. Epstein. 1992. *Beyond Spinoff: Military and Commercial Technologies in a Changing World.* Boston, MA: Harvard Business School Press.

Arrow, Kenneth. 1962. "Economic Welfare and the Allocation of Resources for Invention." Pp. 609 625 in *The Rate and Direction of Inventive Activity: Economic and Social Factors.* A Report of the National Bureau of Economic Research. 609-25. Princeton, NJ: Princeton University Press.

Arrow, Kenneth. 1973. "The Theory of Discrimination." Pp. 3-31 in *Discrimination in Labor Markets.* Orley Ashenfelter and Albert Rees, eds. Princeton, NJ: Princeton University Press.

Audretsch, David B. 1995. *Innovation and Industry Evolution.* Cambridge, MA: MIT Press.

Audretsch, David B. and Maryann P. Feldman. 1996. "R&D Spillovers and the Geography of Innovation and Production." *American Economic Review* 86(3):630–640.

Audretsch, David B. and Paula E. Stephan. 1996. "Company-scientist Locational Links: The Case of Biotechnology." *American Economic Review* 86(3):641–642.

Audretsch, D. and R. Thurik. 1999. *Innovation, Industry Evolution, and Employment.* Cambridge, MA: MIT Press.

Barfield, C. and W. Schambra, eds., 1986. *The Politics of Industrial Policy.* Washington, D.C.: American Enterprise Institute for Public Policy Research.

Baron, Jonathan. 1998. "DoD SBIR/STTR Program Manager." Comments at the Methodology Workshop on the Assessment of Current SBIR Program Initiatives. Washington, D.C. October.

Barry, C. B. 1994. "New Directions in Research on Venture Capital Finance." *Financial Management* 23 (Autumn):3–15.

Bator, Francis. 1958. "The Anatomy of Market Failure." *Quarterly Journal of Economics* 72: 351–379.

Bingham, R. 1998. *Industrial Policy American Style: From Hamilton to HDTV.* New York: M.E. Sharpe.

Birch, D. 1981. "Who Creates Jobs." *The Public Interest* 65(Fall):3–14.

Branscomb, Lewis M. and Philip E. Auerswald. 2001. *Taking Technical Risks: How Innovators, Managers, and Investors Manage Risk in High-Tech Innovations.* Cambridge, MA: The MIT Press.

Branscomb, Lewis M. and Philip Aurswald. 2002. *Between Invention and Innovation: An Analysis of Funding for Early-Stage Technology Development.* NIST GCR 02-841. Prepared for the Economic Assessment Office, Advanced Technology Program. Gaithersburg, MD: National Institute of Standards and Technology. November.

Branscomb, Lewis M. and J. Keller. 1998. *Investing in Innovation: Creating a Research and Innovation Policy.* Cambridge, MA: The MIT Press.

Branscomb, Lewis M., Kenneth P. Morse, Michael J. Roberts, and Darin Boville. 2000. *Managing Technical Risk: Understanding Private Sector Decision Making on Early Stage Technology Based Projects.* Washington, D.C.: Department of Commerce/National Institute of Standards and Technology.

Brav, A. and P. A. Gompers. 1997. "Myth or Reality?: Long-run Underperformance of Initial Public Offerings; Evidence from Venture Capital and Nonventure Capital-backed IPOs." *Journal of Finance* 52:1791–1821.

Brown, G. and J. Turner, 1999. "Reworking the Federal Role in Small Business Research." *Issues in Science and Technology* XV(4, Summer).

Bush, Vannevar. 1946. *Science: the Endless Frontier.* Republished in 1960. Washington, D.C.: U.S. National Science Foundation.

Caves, Richard E. 1998. "Industrial Organization and New Findings on the Turnover and Mobility of Firms." *Journal of Economic Literature* 36(4):1947–1982.

Clinton, William Jefferson. 1994. *Economic Report of the President.* Washington, D.C.: U.S. Government Printing Office.

Coburn, C. and Bergland, D. 1995. *Partnerships: A Compendium of State and Federal Cooperative Technology Programs.* Columbus, OH: Battelle.

Cohen, L. R. and R. G. Noll. 1991. *The Technology Pork Barrel.* Washington, D.C.: The Brookings Institution.

Council of Economic Advisers. 1995. *Supporting Research and Development to Promote Economic Growth: The Federal Government's Role.* Washington, D.C.

Davis, S. J., J. Haltiwanger, and S. Schuh. 1994. "Small Business and Job Creation: Dissecting the Myth and Reassessing the Facts." *Business Economics* 29(3):113–122.

Deputy Under Secretary of Defense (Industrial Policy). 2003. "Transforming the Defense Industrial Base: A Roadmap." February. Accessed at <*http://www.acq.osd.mil/ip/docs/transforming_the_defense_ind_base-full_report_with_appendices.*pdf>.

Dertouzos, M. 1989. *Made in America: The MIT Commission on Industrial Productivity.* Cambridge, MA: The MIT Press.

Eckstein, 1984. *DRI Report on U.S. Manufacturing Industries,* New York: McGraw Hill.

Ehlers, Vernon J. 1998. *Unlocking Our Future: Toward a New National Science Policy: A Report to Congress by the House Committee on Science.* Washington, D.C.: Government Printing Office.

Eisinger, P. K. 1988. *The Rise of the Entrepreneurial State: State and Local Economic Development Policy in the United States.* Madison, WI: University of Wisconsin Press.

Executive Office of the President. 1990. *U.S. Technology Policy.* Washington, D.C.: Executive Office of the President.

Feldman, Maryann P. 1994. *The Geography of Knowledge.* Boston: Kluwer Academic.

Feldman, Maryann P. 1994. "Knowledge Complementarity and Innovation." *Small Business Economics* 6(5):363–372.

Fenn, G. W., N. Liang, and S. Prowse. 1995. *The Economics of the Private Equity Market.* Washington, D.C.: Board of Governors of the Federal Reserve System.

Finley, James I. 2006. Comments at the Small Business Technology Coalition Conference. Washington, D.C. September 27.

Flamm, K. 1988. *Creating the Computer*. Washington, D.C.: The Brookings Institution.

Flender, J. O. and R. S. Morse. 1975. *The Role of New Technical Enterprise in the U.S. Economy*. Cambridge, MA: MIT Development Foundation.

Freear, J., and W. E. Wetzel, Jr. 1990. "Who Bankrolls High-tech Entrepreneurs?" *Journal of Business Venturing* 5:77–89.

Freeman, Chris and Luc Soete. 1997. *The Economics of Industrial Innovation*. Cambridge, MA: The MIT Press.

Galbraith, J. K. 1957. *The New Industrial State*. Boston, MA: Houghton Mifflin.

Geroski, Paul A. 1995. "What Do We Know about Entry?" *International Journal of Industrial Organization* 13(4):421–440.

Gompers, P. A. 1995. "Optimal Investment, Monitoring, and the Staging of Venture Capital." *Journal of Finance* 50:1461–1489.

Gompers, P. A. and J. Lerner. 1996. "The Use of Covenants: An Empirical Analysis of Venture Partnership Agreements." *Journal of Law and Economics* 39:463–498.

Gompers, P. A. and J. Lerner. 1998. "Capital Formation and Investment in Venture Markets: A Report to the NBER and the Advanced Technology Program." Unpublished working paper. Harvard University.

Gompers, P. A. and J. Lerner. 1998. "What Drives Venture Capital Fund-raising?" Unpublished working paper. Harvard University.

Gompers, P. A. and J. Lerner. 1999. "An Analysis of Compensation in the U.S. Venture Capital Partnership." *Journal of Financial Economics* 51(1):3–7.

Gompers, P. A. and J. Lerner. 1999. *The Venture Cycle*. Cambridge, MA: The MIT Press.

Good, M. L. 1995. Prepared testimony before the Senate Commerce, Science, and Transportation Committee, Subcommittee on Science, Technology, and Space Photocopy. U.S. Department of Commerce.

Graham, O. L. 1992. *Losing Time: The Industrial Policy Debate*. Cambridge, MA: Harvard University Press.

Greenwald, B. C., J. E. Stiglitz, and A. Weiss. 1984. "Information Imperfections in the Capital Market and Macroeconomic Fluctuations." *American Economic Review Papers and Proceedings* 74:194–199.

Griliches, Z. 1990. *The Search for R&D Spillovers*. Cambridge, MA: Harvard University Press.

Hall, Bronwyn H. 1992. "Investment and Research and Development: Does the Source of Financing Matter?" Working Paper No. 92–194. Department of Economics, University of California at Berkeley.

Hall, Bronwyn H. 1993. "Industrial Research During the 1980s: Did the Rate of Return Fall?" Brookings Papers. *Microeconomics* 2:289–343.

Hamberg, Dan. 1963. "Invention in the Industrial Research Laboratory." *Journal of Political Economy* (April):95–115.

Hao, K. Y. and A. B. Jaffe. 1993. "Effect of Liquidity on Firms' R&D Spending." *Economics of Innovation and New Technology* 2:275–282.

Hebert, Robert F. and Albert N. Link. 1989. "In Search of the Meaning of Entrepreneurship." *Small Business Economics* 1(1):39–49.

Himmelberg, C. P. and B. C. Petersen. 1994. "R&D and Internal Finance: A Panel Study of Small Firms in High-tech Industries." *Review of Economics and Statistics* 76:38–51.

Hubbard, R. G. 1998. "Capital-market Imperfections and Investment." *Journal of Economic Literature* 36:193–225.

Huntsman, B. and J. P. Hoban, Jr. 1980. "Investment in New Enterprise: Some Empirical Observations on Risk, Return, and Market Structure." *Financial Management* 9(Summer):44–51.

Jaffe, A. B. 1996. "Economic Analysis of Research Spillovers: Implications for the Advanced Technology Program." Gaithersburg, MD: National Institute of Standards and Technology.

Jaffe, A. B. 1998. "The Importance of 'Spillovers' in the Policy Mission of the Advanced Technology Program." *Journal of Technology Transfer* (Summer).

Jewkes, J., D. Sawers, and R. Stillerman. 1958. *The Sources of Invention.* New York: St. Martin's Press.

Kleinman, D. L. 1995. *Politics on the Endless Frontier: Postwar Research Policy in the United States.* Durham, NC: Duke University Press.

Kortum, Samuel and Josh Lerner. 1998. "Does Venture Capital Spur Innovation?" NBER Working Papers 6846. National Bureau of Economic Research.

Krugman, P. 1990. *Rethinking International Trade.* Cambridge, MA: The MIT Press.

Krugman, P. 1991. *Geography and Trade.* Cambridge, MA: The MIT Press.

Langlois, Richard N. and Paul L. Robertson 1996. "Stop Crying over Spilt Knowledge: A Critical Look at the Theory of Spillovers and Technical Change." Paper prepared for the MERIT Conference on Innovation, Evolution, and Technology. Maastricht, Netherlands. August 25–27.

Lebow, I. 1995. *Information Highways and Byways: From the Telegraph to the 21st Century.* New York: Institute of Electrical and Electronic Engineering.

Lerner, J. 1994. "The Syndication of Venture Capital Investments." *Financial Management* 23(Autumn):16–27.

Lerner, J. 1995. "Venture Capital and the Oversight of Private Firms." *Journal of Finance* 50:301–318.

Lerner, J. 1996. "The Government as Venture Capitalist: The Long-run Effects of the SBIR Program." NBER Working Paper 5753. National Bureau of Economic Research.

Lerner, J. 1998. "Angel Financing and Public Policy: An Overview." *Journal of Banking and Finance* 22(6–8):773–784.

Lerner, J. 1999. "The Government as Venture Capitalist: The Long-run Effects of the SBIR Program." *Journal of Business* 72(3):285–297.

Lerner, J. 1999. "Public Venture Capital: Rationales and Evaluation." In National Research Council. *The Small Business Innovation Research Program: Challenges and Opportunities.* Charles W. Wessner, ed. Washington, D.C.: National Academy Press.

Lerner, J. 2000. "Evaluating the Small Business Innovation Research Program: A Literature Review." In National Research Council. *The Small Business Innovation Research Program: An Assessment of the Department of Defense Fast Track Initiative.* Charles W. Wessner, ed. Washington, D.C.: National Academy Press.

Levy, D. M. and N. Terleckyk, 1983. "Effects of Government R&D on Private R&D Investment and Productivity: A Macroeconomic Analysis." *Bell Journal of Economics* 14:551–561.

Liles, P. 1977. *Sustaining the Venture Capital Firm.* Cambridge, MA: Management Analysis Center.

Link, Albert N. 1998. "Public/Private Partnerships as a Tool to Support Industrial R&D: Experiences in the United States." Paper prepared for the Working Group on Innovation and Technology Policy of the OECD Committee for Science and Technology Policy. Paris, France: Organisation for Economic Co-operation and Development.

Link, Albert N. and John Rees. 1990. "Firm Size, University-based Research and the Returns to R&D." *Small Business Economics* 2(1):25–32.

Link, Albert N. and John T. Scott. 1998. "Assessing the Infrastructural Needs of a Technology-based Service Sector: A New Approach to Technology Policy Planning." *STI Review* 22:171–207.

Link, Albert N. and John T. Scott. 1998. *Overcoming Market Failure: A Case Study of the ATP Focused Program on Technologies for the Integration of Manufacturing Applications (TIMA).* Draft Final Report Submitted to the Advanced Technology Program. Gaithersburg, MD: National Institute of Technology. October.

Link, Albert N. and John T. Scott. 1998. *Public Accountability: Evaluating Technology-Based Institutions.* Norwell, MA.: Kluwer Academic.

Malone, T. 1995. *The Microprocessor: A Biography.* Hamburg, Germany: Springer Verlag/Telos.

Mansfield, E. 1985. "How Fast Does New Industrial Technology Leak Out?" *Journal of Industrial Economics* 34(2).

Mansfield, E. 1996. *Estimating Social and Private Returns from Innovations Based on the Advanced Technology Program: Problems and Opportunities.* Unpublished report.

Mansfield, E., J. Rapoport, A. Romeo, S. Wagner, and G. Beardsley. 1977. "Social and Private Rates of Return from Industrial Innovations" *Quarterly Journal of Economics* 91:221–240.

Martin, Justin. 2002. "David Birch." *Fortune Small Business.* December 1.

McCraw, T. 1986. "Mercantilism and the Market: Antecedents of American Industrial Policy." In C. Barfield and W. Schambra, eds., *The Politics of Industrial Policy.* Washington, D.C.: American Enterprise Institute for Public Policy Research.

Mervis, Jeffrey D. 1996. "A $1 Billion 'Tax' on R&D Funds." *Science* 272:942–944.

Mowery, D. 1998. "Collaborative R&D: How Effective Is It." *Issues in Science and Technology.* (Fall):37–44.

Mowery, D., ed. 1999. *U.S. Industry in 2000: Studies in Competitive Performance.* Washington, D.C.: National Academy Press.

Mowery, D. and N. Rosenberg. 1989. *Technology and the Pursuit of Economic Growth.* New York: Cambridge University Press.

Mowery, D. and N. Rosenberg. 1998. *Paths of Innovation: Technological Change in 20th Century America.* New York: Cambridge University Press.

Myers, S., R. L. Stern, and M. L. Rorke. 1983. *A Study of the Small Business Innovation Research Program.* Lake Forest, IL: Mohawk Research Corporation.

Myers, S. C. and N. Majluf. 1984. "Corporate Financing and Investment Decisions When Firms Have Information that Investors Do Not Have." *Journal of Financial Economics* 13:187–221.

National Research Council. 1986. *The Positive Sum Strategy: Harnessing Technology for Economic Growth.* Washington, D.C.: National Academy Press.

National Research Council. 1987. *Semiconductor Industry and the National Laboratories: Part of a National Strategy.* Washington, D.C.: National Academy Press.

National Research Council 1991. *Mathematical Sciences, Technology, and Economic Competitiveness.* James, G. Glimm, ed. Washington, D.C.: National Academy Press.

National Research Council. 1995. *Allocating Federal Funds for R&D* Washington, D.C.: National Academy Press.

National Research Council. 1996. *Conflict and Cooperation in National Competition for High-Technology Industry.* Washington, D.C.: National Academy Press.

National Research Council. 1997. *Review of the Research Program of the Partnership for a New Generation of Vehicles: Third Report.* Washington, D.C.: National Academy Press.

National Research Council. 1999. *The Advanced Technology Program: Challenges and Opportunities.* Charles W. Wessner, ed. Washington, D.C.: National Academy Press.

National Research Council. 1999. *Funding a Revolution: Government Support for Computing Research.* Washington, D.C.: National Academy Press.

National Research Council. 1999. *Industry-Laboratory Partnerships: A Review of the Sandia Science and Technology Park Initiative.* Charles W. Wessner, ed. Washington, D.C.: National Academy Press.

National Research Council. 1999. *New Vistas in Transatlantic Science and Technology Cooperation.* Charles W. Wessner, ed. Washington, D.C.: National Academy Press.

National Research Council. 1999. *The Small Business Innovation Research Program: Challenges and Opportunities.* Charles W. Wessner, ed. Washington, D.C.: National Academy Press.

National Research Council. 2000. *The Small Business Innovation Research Program: A Review of the Department of Defense Fast Track Initiative.* Charles W. Wessner, ed. Washington, D.C.: National Academy Press.

National Research Council. 2000. *U.S. Industry in 2000: Studies in Competitive Performance.* Washington, D.C.: National Academy Press.

National Research Council. 2001. *The Advanced Technology Program: Assessing Outcomes.* Charles W. Wessner, ed. Washington, D.C.: National Academy Press.

National Research Council. 2001. *Attracting Science and Mathematics Ph.Ds to Secondary School Education.* Washington, D.C.: National Academy Press.

National Research Council. 2001. *Building a Workforce for the Information Economy.* Washington, D.C.: National Academy Press.

National Research Council. 2001. *Capitalizing on New Needs and New Opportunities: Government-Industry Partnerships in Biotechnology and Information Technologies.* Charles W. Wessner, ed. Washington, D.C.: National Academy Press.

National Research Council. 2001. *A Review of the New Initiatives at the NASA Ames Research Center.* Charles W. Wessner, ed. Washington, D.C.: National Academy Press.

National Research Council. 2001. *Trends in Federal Support of Research and Graduate Education.* Stephen A. Merrill, ed. Washington, D.C.: National Academy Press.

National Research Council. 2002. *Government-Industry Partnerships for the Development of New Technologies: Summary Report.* Charles W. Wessner, ed. Washington, D.C.: The National Academies Press.

National Research Council. 2002. *Measuring and Sustaining the New Economy.* Dale W. Jorgenson and Charles W. Wessner, eds. Washington, D.C.: National Academy Press.

National Research Council. 2004. *The Small Business Innovation Research Program: Program Diversity and Assessment Challenges.* Charles W. Wessner, ed. Washington, D.C.: The National Academies Press.

National Research Council. 2005. *Assessment of Department of Defense Basic Research.* Washington, D.C.: The National Academies Press.

Nelson, R. R. 1982. *Government and Technological Progress.* New York, NY: Pergamon.

Nelson, R. R. 1986. "Institutions Supporting Technical Advances in Industry." *American Economic Review: Papers and Proceedings* 76(2):188.

Nelson, R. R., ed. 1993. *National Innovation System: A Comparative Study.* New York, NY: Oxford University Press.

The New York Times. 2005. "Pentagon Redirects Its Research Dollars." April 2.

Office of Management and Budget. 1996. *Economic Analysis of Federal Regulations Under Executive Order 12866.* Mimeo.

Organisation for Economic Co-operation and Development. 1982. *Innovation in Small and Medium Firms.* Paris, France: Organisation for Economic Co-operation and Development.

Organisation for Economic Co-operation and Development. 1995. *Venture Capital in OECD countries.* Paris, France: Organisation for Economic Co-operation and Development.

Organisation for Economic Co-operation and Development, 1997. *Small Business Job Creation and Growth: Facts, Obstacles, and Best Practices.* Paris, France: Organisation for Economic Co-operation and Development.

Organisation for Economic Co-operation and Development. 1998. *Technology, Productivity and Job Creation: Toward Best Policy Practice.* Paris, France: Organisation for Economic Co-operation and Development.

Perret, G. 1989. *A Country Made by War: From the Revolution to Vietnam—The Story of America's Rise to Power.* New York, NY: Random House.

Powell, Walter W., and Peter Brantley. 1992. "Competitive Cooperation in Biotechnology: Learning Through Networks?" Pp. 366–394 in N. Nohria and R. G. Eccles, eds., *Networks and Organizations: Structure, Form and Action.* Boston, MA: Harvard Business School Press.

Price Waterhouse. 1985. *Survey of Small High-tech Businesses Shows Federal SBIR Awards Spurring Job Growth, Commercial sales.* Washington, D.C.: Small Business High Technology Institute.

Roberts, Edward B. 1968. "Entrepreneurship and Technology." *Research Management* (July): 249-266.

Romer, P. 1990. "Endogenous Technological Change." *Journal of Political Economy* 98:71–102.

Rosenbloom, R., and Spencer, W. 1996. *Engines of Innovation: U.S. Industrial Research at the End of an Era.* Boston, MA: Harvard Business Press.

Rubenstein, A. H. 1958. *Problems Financing New Research-Based Enterprises in New England.* Boston, MA: Federal Reserve Bank.

Sahlman, W. A. 1990. "The Structure and Governance of Venture Capital Organizations." *Journal of Financial Economics* 27:473–521.

Saxenian, Annalee. 1994. *Regional Advantage: Culture and Competition in Silicon Valley and Route 128.* Cambridge, MA: Harvard University Press.

Scherer, F. M. 1970. *Industrial Market Structure and Economic Performance.* New York: Rand McNally College Publishing.

Schumpeter, J. 1950. *Capitalism, Socialism, and Democracy.* New York: Harper and Row.

Science. 2005. "An Endless Frontier Postponed." May 6.

Scott, John T. 1998. "Financing and Leveraging Public/Private Partnerships: The Hurdle-lowering Auction." *STI Review* 23:67–84.

Sohl, Jeffery, John Freear, and W. E. Wetzel, Jr. 2002. "Angles on Angels: Financing Technology-Based Ventures–An Historical Perspective." *Venture Capital: An International Journal of Entrepreneurial Finance* 4(4).275-287.

Spence, Michael. 1974. *Market Signaling: Informational Transfer in Hiring and Related Processes.* Cambridge, MA: Harvard University Press.

Stiglitz, J. E. and A. Weiss. 1981. "Credit Rationing in Markets with Incomplete Information." *American Economic Review* 71:393–409.

Stowsky, J. 1996. "Politics and Policy: The Technology Reinvestment Program and the Dilemmas of Dual Use." Mimeo: University of California.

Tassey, Gregory. 1997. *The Economics of R&D Policy.* Westport, CT: Quorum Books.

Tirman, John. 1984. *The Militarization of High Technology.* Cambridge, MA: Ballinger.

Tyson, Laura, Tea Petrin, and Halsey Rogers. 1994. "Promoting Entrepreneurship in Eastern Europe." *Small Business Economics* 6:165–184

U.S. Congress, Senate Committee on Small Business. 1981. *Small Business Research Act of 1981.* S.R. 194, 97th Congress.

U.S. Congressional Budget Office. 1985. *Federal Financial Support for High-technology Industries.* Washington, D.C.: U.S. Congressional Budget Office.

U.S. General Accounting Office. 1987. *Federal Research: Small Business Innovation Research Participants Give Program High Marks.* Washington, D.C.: U.S. General Accounting Office.

U.S. General Accounting Office. 1989. *Federal Research: Assessment of Small Business Innovation Research Program.* Washington, D.C.: U.S. General Accounting Office.

U.S. General Accounting Office. 1992. *Federal Research: Small Business Innovation Research Shows Success But Can be Strengthened.* RCED 92-32. Washington, D.C.: U.S. General Accounting Office.

U.S. General Accounting Office. 1997. *Federal Research: DoD's Small Business Innovation Research Program.* RCED–97–122. Washington, D.C.: U.S. General Accounting Office.

U. S. General Accounting Office. 1998. *Federal Research: Observations on the Small Business Innovation Research Program.* RCED–98–132. Washington, D.C.: U.S. General Accounting Office.

U.S. General Accounting Office. 1999. *Federal Research: Evaluations of Small Business Innovation Research Can Be Strengthened.* RCED–99–198. Washington, D.C.: U.S. General Accounting Office.

U.S. House of Representatives Committee on Science. 1998. *Unlocking Our Future: Toward a New National Science Policy—A Report to Congress by the House Committee on Science.* Washington, D.C.: Government Printing Office.

U.S. Senate Committee on Small Business. 1981. Senate Report 97–194. *Small Business Research Act of 1981.* September 25. Washington, D.C.: U.S. Government Printing Office.

U.S. Senate Committee on Small Business. 1999. Senate Report 106–330. *Small Business Innovation Research (SBIR) Program*. August 4. Washington, D.C.: U.S. Government Printing Office.

U.S. Small Business Administration. 1992. *Results of Three-Year Commercialization Study of the SBIR Program*. Washington, D.C.: U.S. Government Printing Office.

U.S. Small Business Administration. 1994. *Small Business Innovation Development Act: Tenth-year Results*. Washington, D.C.: U.S. Government Printing Office.

U.S. Small Business Administration. 2003. "Small Business by the Numbers." SBA Office of Advocacy. May.

U.S. Small Business Administration. 2006. TechNet Data Base. <*http://tech-net.sba.gov/*>. Accessed on July 25, 2006.

Venture Economics. 1988. *Exiting Venture Capital Investments*. Wellesley, MA: Venture Economics.

Venture Economics. 1996. "Special Report: Rose-colored Asset Class." *Venture Capital Journal* 36(July):32–34 (and earlier years).

VentureOne. 1997. *National Venture Capital Association 1996 Annual Report*. San Francisco: VentureOne.

The Wall Street Journal. 2006. "The Venture Capital Yard Sale." July 18. Page C1.

Wallsten, S. J. 1996. "The Small Business Innovation Research Program: Encouraging Technological Innovation and Commercialization in Small Firms." Unpublished working paper, Stanford University.